D0904809

Patterns in Nature

Patterns in Nature

The Analysis of Species Co-occurrences

JAMES G. SANDERSON
STUART L. PIMM

THE UNIVERSITY OF CHICAGO PRESS   CHICAGO AND LONDON

JAMES G. SANDERSON is a member of the IUCN Cat Specialist Group, fellow of the Wild-
life Conservation Network, and founder and director of the Small Wild Cat Conservation
Foundation. STUART L. PIMM is the Doris Duke Professor of Conservation Ecology at Duke
University.

The University of Chicago Press, Chicago 60637
The University of Chicago Press, Ltd., London
© 2015 by The University of Chicago
All rights reserved. Published 2015.
Printed in the United States of America

24 23 22 21 20 19 18 17 16 15   1 2 3 4 5

ISBN-13: 978-0-226-29272-4 (cloth)
ISBN-13: 978-0-226-29286-1 (e-book)
DOI: 10.7208/chicago/9780226292861.001.0001

Library of Congress Cataloging-in-Publication Data
Sanderson, James G., 1949–author.
    Patterns in nature : the analysis of species co-occurrences / James G. Sanderson and
Stuart L. Pimm.
        pages ; cm
    Includes bibliographical references and index.
    ISBN 978-0-226-29272-4 (cloth : alk. paper)—ISBN 978-0-226-29286-1 (e-book)
    1. Biogeography. 2. Biotic communities. 3. Null models (Ecology) 4. Pattern formation
(Biology) I. Pimm, Stuart L. (Stuart Leonard), author. II. Title.
    QH84.S26 2015
    577.2'2—dc23

                                    2015011314

♾ This paper meets the requirements of ANSI/NISO Z39.48-1992 (Permanence of Paper).

4367947↑

Contents

Preface

Robert MacArthur died in 1972. He is one of the most frequently cited ecologists. With Ed Wilson, he coauthored the book *Island Biogeography*, which may be the most highly cited publication in ecology ever. Yet MacArthur died at only forty-two, so his remarkable influence comes from a tragically short career. He published his first paper in 1955. Some of the most influential ecologists who gathered to celebrate his memory published their memorial volume in 1975. Why was his legacy so powerful?

We are neither historians nor philosophers of science, nor is this book our attempt to become such. Rather, we are interested in the patterns of nature. Jared Diamond wrote by far the longest chapter in the memorial volume, and it seems to us to capture the essence of MacArthur's passion. Diamond's chapter is frequently cited to this day.

Diamond's paper dealt with the distribution of birds on different islands and, in some cases, how bird species divide up islands where they live together. These patterns are large scale, involve many species, and surely take a long time to develop. As we explain in the first chapter, these patterns are at the core of the simplest observations we make as naturalists. That research space is relatively only sparsely populated, yet hugely important. One of us has said all this before (Pimm 1991), so we will not belabor the point, here.

What makes this book's subject especially interesting is the reaction it caused. Within a few years, Dan Simberloff—one of Wilson's students—with one of his own students published a scathing attack on Diamond's work. Simberloff, too, is a scientist of considerable influence.

Some of what happened thereafter was simply ugly, involving one of the most violent clashes of personalities in recent ecological history. Putman (1994) remarked that the debate was "almost unprecedented in the apparent entrenchment and hostility of the opposing camps." This debate was not ecology's finest hour. Even now, thirty years later, emotionally charged manuscripts are still appearing. We can also say that in our careers, we have had no experiences in which the traditions of scientific publication have been so consistently ignored. As one example, in the latest salvo in this debate, Connor et al. (2013) criticizing Sanderson et al. (2009), the journal editor did not send the manuscript for comment to Sanderson and his colleagues. Such is the common practice, even as editors understand that those criticized may hardly be disinterested parties. The practice can prevent trivial misunderstandings, however. As we show, Connor et al. (2013) made a trivial error that invalidates their paper's conclusions, something that even a most cursory review would have spotted.

It is unlikely this book will bring closure. Our goals are more modest. First, we modernize the null model methods and move them from a theoretical framework to a powerful tool to investigate problems in community ecology. To achieve our goals we take a historical perspective that began at Princeton University at MacArthur's eulogy and subsequent 1975 publication of Diamond's famous chapter on assembly rules. We thus summarize Diamond's work to launch our presentation and follow with Connor and Simberloff's (1979) critique of assembly rules and community ecology. With the benefit of hindsight, we show that both sides had good ideas worth pursuing, while both suffered from weaknesses that the opposing side continually exploited. We are realistic and accept that closing the vast neutral zone between the two opposing camps is likely impossible, in spite of the fact that we regard the analysis of species co-occurrences to be a solved problem. For the record, we have coauthored several papers with Diamond, and Sanderson has coauthored a paper with Simberloff.

We do our best here to present the intellectual content of the arguments. The reason to write this book is not to unearth bitterness. Rather, our second goal is to show that the intellectual issues are substantial and important. They span a period of dramatic growth in computer power, without which modern analysis would have been impossible. Even more important have been the significant conceptual issues of building the models to describe communities. Above all, this is about the search for large-scale patterns in nature. That is where we start.

Acknowledgments

We thank John Faarborg, Clinton Jenkins, Michael Moulton, Michael Rosenzweig, and two anonymous reviewers for their comments. Jenkins also provided substantial help with mapping, while Doug Pratt allowed us to use his drawings to illustrate some of the species we consider. Josep del Hoyo and *Handbook of the Birds of the World* graciously allowed us to create figures with material from that book.

PART I

The Distribution of Species on Islands

Patterns or Fantasies?

Species Co-occurrences

This book is about the identification and interpretation of nature's large-scale patterns of species co-occurrence and what we can deduce from them about how nature works. Surely, the earliest experiences we have as naturalists are putting names to the species we see. Jim's first memory was of a white-tailed deer, Stuart's of a goldfinch and a reed bunting. Jim knew his deer was not a mule deer because he saw it in New Jersey and Stuart that his bunting was not a little bunting because he saw it in England. Thereafter, across the planet, on countless occasions we have reached for field guides to check identifications. Certainly, some species are easy, colorful, loud, or otherwise distinctive—like the goldfinch. Other groups of species are difficult, and the field guides help by providing maps adjacent to the illustrations, for many similar species live in different places. So, geographical range helps—and often helps a lot. So, too, does habitat. Field guides frequently describe habitat choices as a means to help identification.

We pass this wisdom on to those we mentor. "Yes, species A looks like species B, but the former is in pine forest usually, the latter in deciduous forests." When bird species are on migration, all bets are off. Migrant warblers are famously difficult in both Old and New Worlds, exactly because clues of distribution and habitat are of little help. In short, patterns of species co-occurrence—which species overlap, and more particularly which do not—are fundamental, ubiquitous, and large-scale patterns of nature. The inescapable question is what causes these patterns.

As a later chapter documents, it was noticing similar but different species of birds on the different islands in the Galápagos and primates on different banks of the Amazon that gave Darwin and Wallace, respectively, their essential insights into evolution. Our question has significant implications for understanding nature.

The Night Sky Effect

The notion of finding patterns is enormously seductive. There is an unmistakable feeling of "Ah Gotcha!" It tells us that, amid the bewildering complexity of the natural world, there is something "out there" that is predictable. There is something we can hope to understand. A central point of this book is that this understanding does not come cheaply.

Look up at the night sky. You see Orion the hunter, the Big Dipper, the Southern Cross, and so on. Depending on where you are in the world, these are immediately obvious, familiar since childhood, and you learn new patterns or teach them to others with delight. "Look, there's Scorpio, you can see the stingers in its tail." Even the fact that Orion is doing cartwheels in the Southern Hemisphere—he's upside down, of course, with his dog flying over him—doesn't spoil the pleasure.

They seem so real; shame on us for suggesting that the astral patterns are an illusion. The stars are scattered about randomly. What seem to us to be associations are unrelated stars, huge distances apart from each other and in their distances from us. We see pattern where there is just randomness.

Wait! On a clear night away from city lights, we can see a dense band of stars stretching from horizon to horizon, spectacular clusters seen through the binoculars. The Milky Way is not random: it is a band spread across the sky. *That pattern means something.* Eventually, we conclude that some process organizes billions of stars into a galactic pancake. A visit to the Southern Hemisphere quickly shows that those who live there have the better seats. The Milky Way is thicker there, brighter than in the north, so likely north faces outward from the pancake's center, south toward the center. Our home star, the sun, must be on the unfashionable outer edge of our galaxy.

Through binoculars, look between a huge open square of stars, Pegasus—though we have never quite made a horse of them—and a *W*, Cassiopeia—a beautiful woman, apparently. (The need to see patterns is a powerful human urge.) Between the square and the *W* there is a small but entire pancake on its own, slightly on edge. After a clever bit of observation—with a telescope, not with binoculars—we know it to be a

galaxy way beyond our own. More observations and we know these gal-
axies are all moving away from each other at prodigious speeds, and the
farther they are from us, the faster they are moving.

Ecology and astronomy have much in common. Ecologists, like astronomers, have no more chance to do neat controlled experiments at relevant
scales in space and time. We can observe patterns, however. Perhaps we
can observe them changing over decades, but that's a period we know is
absurdly short given how long the universe—or life on Earth—has been
around. Physicists can do experiments of a kind—they can swing their pen-
dulums and make their measurements over much the same small scales.
Ecologists can certainly do experiments, too, but these experiments are
typically across just a few square meters and over a small number of years
(Pimm 1991).

Like astronomers, ecologists must embrace what we can only infer
from observations at scales that are far too large for us to tinker. And both
professions must connect these observations to small-scale and hopefully
relevant experiments that test the processes that might operate at those
large scales.

Patterns in Nature

In the late 1960s and early 1970s, MacArthur, at Princeton University, led
ecologists in studying the large-scale patterns of nature. He envisioned
ecologists making field observations and, from the patterns they deduced,
creating theories that explained them. Such theories would enable predic-
tion of new and novel facts observed at places yet unstudied. From theories arose testable hypotheses.

MacArthur looked at patterns involving many species. He was not
alone. In 1966, the famous British ecologist Charles Elton published *The
Pattern of Animal Communities*. Five years later David Lack, also at Oxford, published *Ecological Isolation in Birds*. All these ecologists were
looking at communities—collections of species, not just single species—
and they were doing so over large spatial scales. What made MacArthur's
work particularly exciting was his ability to synthesize general patterns
and to offer crisp theories to explain them. The ideas had great vitality,
and they prompted many questions. Community ecology is the ecology
of collections of species, and the most obvious question one can ask is
how large is the collection? That is, does a community have few species or
many? Such questions stand in stark contrast to other fields of ecology—

such as how and why the numbers of a particular species vary from species to species, place to place, and year to year.

Today there are millions of bird-watchers worldwide, and they keep lists of which species they have seen and where and when they saw them. Bragging rights go to those who have seen a species in some place or time where or when it should not have been, the excitement being so intense that one poor chap would twitch when he saw something new. "Twitchers"—the most passionate birders—nonetheless collect fundamental ecological data on bird communities. So, too, do listers of other groups of species, though generally not with the same fanaticism.

What occurs where (and why), and why some places harbor more species than others are basic questions for ecologists and listers alike. Part of the answer—probably the large part and for some the only part—is that species simply live in different places. There are countless examples of such patterns. Some are trivial. Fish live underwater. Birds do not. Adaptations follow: most fish have gills; birds have lungs. "A fish out of water" is the expression for a person—and an animal—in the wrong place.

Two things should nag us. The first is that not all the patterns are so seemingly trivial. Travel along any gradient—up a mountain, from forest into desert, from a north-facing slope to a south-facing slope, from low tide to high tide on a shoreline, from Arctic tundra to tropical rain forest—and the species change. It does not matter what species you are studying. Why does one species end exactly *here* and another, *there*?

Certainly, the answer may be that the patterns are random. What *random* means is hugely controversial. It could mean "without cause," implying that there is some irreducible uncertainty in the universe. It could also mean "there are too many causes for us to understand what is happening." A species range may end *over there* because beyond *over there* it is too hot, too dry, too high, too dark—or too whatever on a very long list of possible factors. Simply, it may be possible that species associations are not random, but there is no way for us to tell.

The most famous random event is tossing a coin. The moment the coin leaves the fingers that have tossed it into the air, its path is entirely predictable from the forces acting upon it. Whether the coin comes up heads or tails is totally determined, but the number and complexity of the many interacting factors hamper our ability to predict the result. The stars in Orion have their striking relationship because of an incomprehensible mess of gravitational forces unfolding over the entire history of the universe.

Before going further, we must define what the word *random* means in our context. We define *random* to mean an event or outcome that does not differ from chance expectations, as, for example, the number of times two species are observed to co-occur on a group of islands.

The question is not whether species associations are "random." Rather, the question is one of recognition: Can we differentiate associations caused by a multiplicity of complex, idiosyncratic factors from those structured by some unidentified, but simple mechanisms? Can we infer simple mechanisms that structure communities from observations of which species associations naturally occur?

Diamond provided an emphatic yes to this question. He had been bird-watching on the Bismarck islands near New Guinea. He did what millions of bird-watchers do around the world as they go into the field: he kept a list of which species he saw and where he saw them.

Diamond's observations were that some pairs of bird species, particularly ecologically similar ones, did not occur together on any islands in an archipelago. He took that to be evidence for the simple mechanism of competitive exclusion. Implicit in this argument was that the islands were, to a first approximation, sufficiently similar to each other, that the myriad of other factors that might explain where one species began and another ended were relatively unimportant.

Diamond (1975) continued, suggesting that island bird communities were structured by *assembly rules* that could be deduced from observation of which birds occurred on particular islands. Diamond championed the idea that a *checkerboard distribution pattern* (i.e., a pattern of mutual exclusivity) was the simplest pattern that might occur under competitive exclusion. That is, if species A occupies an island, then species B cannot because A prevents B from colonizing the island and vice versa. Apparently A and B are so similar ecologically that only one can occupy a given island. Which species got there first was an accident.

Importantly, the ideas generalize to all patterns of species occurrence. What makes islands important—and these islands in particular—was that one could see the patterns here so clearly. Islands influenced Darwin and Wallace so powerfully for the same reason: islands provided clear examples of a process—evolution in their case—that was nonetheless ubiquitous.

The night sky analogy is relevant again. We can easily see patterns— indeed, we seem to be powerfully seduced by the need to see them—but that does not mean they are real. Subtle patterns, however, may betray the most powerful forces that shape nature at her largest scales. How can we be sure that checkerboards are "real"?

Now consider the second question that should nag us. Could it be that because the mechanisms had already acted to create communities, there is simply no way of elucidating the difference between random associations of communities and those structured by the important mechanism?

It is not trivial that birds live on land and fish live in water. Birds came from reptiles whose "killer app"—their key adaptation—was the amniote egg—an egg they did not need to lay in water. Reptiles came from amphibians, which generally must lay their eggs in water, even though the adults can live on land. Is this because some amphibians survived better when on land, by avoiding competition or predation from fish? Competition explains why we are what we are today (including some of the fine details, according to those who study recent hominid fossil history), even if we do not now compete with fish.

Finding the Null

We could mean many things by the phrase "patterns in nature." When faced with seemingly overwhelming complexity in even the simplest natural communities, perhaps asking a simple question is best: Does a pair of species occur together more or less than expected by chance?

The so-called null matrix problem was born in an ecological context in 1979 when Connor and Simberloff (1979) reviewed Diamond's 1975 work. They claimed that the observed numbers of pairs of bird species on the Vanuatu (formerly the New Hebrides), an archipelago east of New Guinea, and the bats and birds of the West Indies closely agreed with what one would expect by chance. That is, the distributions are random.

Connor and Simberloff never claimed that members of a community were fundamentally randomly distributed. Rather, they argued there was no way one could tell the difference between a random assembly and one structured by some unknown mechanism. Diamond was saying, "look, I see patterns," to which Connor and Simberloff replied, "no, it is just your imagination." From those patterns, Diamond deduced that competition structured the patterns of species occurrences at large geographical scales. Connor and Simberloff would have none of that.

From his notebook, Diamond could create a table—a matrix. Rows are the list of species found across the archipelago and columns the individual islands. A 1 means that you saw that species on that island; a 0 means that you did not. A checkerboard looks like this:

	Islands				
Species	A	B	C	D	F
Golden-headed ant-tender	0	0	1	1	1
Lesser spotted gnat-impaler	1	1	0	0	0

To derive their conclusions, Connor and Simberloff (1979) created a collection of random, or null, communities. They called this the *sample null space*. For both the observed *incidence matrix*—the matrix of 0's and 1's derived from the observed community—and each random community, they calculated the values of a metric used to describe the patterns.

We apologize for the jargon. The incidence matrix is simply a table of which species occur where. Jim has his notes of which cats he saw where, Stuart of which birds he saw where. The null space is a collection of matrices that resemble the observed matrix in important ways, but we chose at random which islands have which species. We call it a "sample" because it is almost never the complete collection of such matrices. Complete collections can have impossibly large numbers of possibilities.

Connor and Simberloff argued that Diamond's assembly rules were either "tautologies" or "trivial consequences of the birds' distributions." They responded to Diamond's chapter by developing a random (or null) model with which to test the difference between an observed community and a collection of random communities. They tested their null model on several communities and concluded that none differed from what was expected were they to be randomly distributed. Again, they did *not* conclude the communities were random. Rather, they concluded that there was no significant difference between the observed community and a collection of randomly placed species. This is a subtle but fundamental difference. But with patterns hard to confirm, many ecologists lost interest in community ecology. Indeed it is difficult to overestimate the impact Connor and Simberloff's (1979) contribution had on community ecology. Since there was no way to distinguish random from structured communities, why pursue studies in community ecology?

For the aforementioned example, notice that there is a 50–50 chance of finding a species on an island—there are five 1's and five 0's. Visit your nearest roulette wheel. Place your first bet on red (the ant-tender absent on island A), the next one on black (the ant-tender present). Then repeat for your next pair of bets. Finally, for the last three pairs of bets, place the first of each pair on black, the second on red. It is no wonder these are called Monte Carlo methods. You could write a computer program that picks 1's and 0's with equal probability. Computers can do this millions of times and never complain.

Now, if you count the number of times you get either a 0 on one island and a 1 on the other—it does not matter which island is which—you will get this checkerboard pattern in about 3% of the attempts. Patterns when the species never occur together on five islands are quite common. They occur about 25% of the time.

We hasten to point out that this is not exactly what Connor and Simberloff did. Still, it helps give a sense of the chaos that followed. The checkerboard you generated using these methods is not special; there is no pattern, you did not win a jackpot, and the journal *Nature* will not publish your paper.

Unfortunately, the outcomes of our Monte Carlo experiment include a large number of cases in which neither species occurs on some islands. In fact, one in a thousand show no species on any of the islands. These outcomes make no ecological sense. The answer, clearly, is that one should build a more sensible set of null models that properly reflect the logical and ecologically sensible constraints.

Unfortunately, building sensible null models turns out to be fiendishly difficult. Why this is the case is something that takes several chapters to explain. It is a story about concepts central to understanding large-scale patterns, technical issues involving null models, and the ability to run those models given the computing power available at the time.

What This Book Is About

This book is about determining whether some species co-occurrences differ from chance expectations. Over the decades that followed, many attacked the null model Connor and Simberloff offered, including Diamond and his colleagues. Yet although the null model was under attack, it was not totally discredited, for no other usurped it. Several authors saw its value and offered improvements. Time has shown that the underlying principles of the null model were solid; the devil was in the details of the model. The method used to create the collection of random communities was complex. Choosing the proper metric by which to statistically test difference between observed and random was like choosing the proper jury to begin a trial. Once the jury is chosen, the trial is half over.

This book is about some of the patterns in nature, or alternatively, about the laws of nature. Many misuse the term *law* in science. A law is no more and no less than a widespread pattern in nature. The characteristic way in which the number of species one finds across islands increases with island size is such a law, even though we often call it the "species area relationship." Other laws are the observations that islands have fewer species on them the farther they are away from a mainland and that, all other things being equal, species with small geographic ranges are rarer within those ranges than are species with large ones. There are several such laws of biological diversity (Pimm and Jenkins 2010). In this sense, some of what we routinely lump into the phrase "the theory of evolution" is a law. It is a collection of facts—for example, that related species appear adjacent to each other in time and space. (One does not find dinosaurs scattered haphazardly in the fossil record, a few in the Cambrian, a few in the Jurassic, the odd one now roaming remote tepuis of Venezuela and nothing in between, for example.) The biogeographer Alfred Russell Wallace assembled those facts of evolution in one of the great triumphs of nineteenth-century science, his astonishing "Sarawak law" paper (Wallace 1855). Two years later, he and Charles Darwin independently came up with the mechanism to explain those laws—the theory of evolution by natural selection.

This book, too, is about laws and their explanations.

It is not about *all* patterns. We do not consider those laws of diversity already mentioned, for instance. Rather, we consider a special, conceptually difficult class of problems that we can reduce to the simple table we illustrated with the presences and absences of species in different places. This might seem restrictive, but it typifies a broad class of problems.

For another example, suppose that as we count the hummingbird species we encounter on our next birding trip to South America, we record what species of flowers they visit. The table of which birds visit which flowers looks just like a table of species on islands. We will ask questions about what patterns emerge from their study.

Not only are there many patterns that fall into this class, but there are also many processes that generate comparable patterns. Most ecological studies of more than one species involve the interactions between a mere two species, even though they embed in a larger community of species. The mechanisms involved in the interaction include competition for resources, predation and predator–prey interactions, mutualism, and exploitation. We explore all these mechanisms in this book, but competition is our starting point and provides most of our examples.

Under natural conditions, these simple direct interactions embed within a community so that other members of the community can alter the strength of the interaction. Physical complexity may also play an important role in determining the outcome of species interactions. Often the role of a top predator or grazer enhances and maintains local diversity. For instance, an herbivore that consumes a plant that shades other plants—and so prevents them from becoming established—may increase local diversity.

Additional competitors of dominant species can have similar effects on communities by effectively suppressing the population growth rate of the dominant competitor. The addition of lesser competitors might thus aid a weaker competitor. We call this *competitive mutualism.*

As we progressively add more species to the mix, the bestiary of possibilities of who does what to whom indirectly becomes ever larger. A wealth of studies, kicked off by the experiments of Connell (1961) and Paine (1966) in intertidal communities, uncovers their existence. An exhaustive review of those experiments by Menge (1995) shows some possibilities more common than are others, with ever more complex patterns, as more species become involved.

Alternatively, the effects of (say) direct competition might be enhanced by the presence of other competitors even if only interacting weakly with the strongest competitors—an effect referred to as *diffuse competition.* Such diffuse competition may have a major contributing effect on the observed response of species in the community (Putman 1994, 38).

In detecting large-scale patterns, much hinges on which species occur where. In the context of birds on islands, one might make the argument that sampling issues come into play, that some birds likely occurred on some islands that were not detected because we do not always detect them even when they are present. Moreover, community composition might change in time. If so, the proportion of islands occupied by a single species must be estimated using occupancy modeling (MacKenzie et al. 2005). Even if this were done, the result would be an estimate of the number of times a species occurs, and not on which islands it might have failed to be detected. This rule would apply to all species, rendering the analysis of species co-occurrences more difficult. Following others, including Diamond and his colleagues, and Simberloff and his colleagues, we chose to accept that all islands were thoroughly sampled and that few birds that might occur on an island or site were missed—so few that no material difference in results would be found.

How This Book Is Organized

In the next chapter, we present Diamond's assembly rules. We do so by setting the historical context first, by introducing MacArthur and the exciting research program he envisioned. We then consider the islands and the birds that Diamond studied, and finally the rules themselves. There is a considerable amount of relevant natural history—Diamond's (1975) book chapter was very long. Even more information has appeared since.

In chapter 3, we present the backlash from Connor and Simberloff. In a series of papers, they argued that Diamond's assembly rules were poorly constructed and that, moreover, his observations did not support them. We agree that, with hindsight, Diamond could have expressed his rules more concisely, but at their core are two powerful assertions about which species occur where. Finally, we take exception to the methods Connor and Simberloff employed. Certainly, they made an important contribution to the study of ecological patterns by requiring observed distributions to be compared to carefully constructed null hypotheses. Their methods, however, needed substantial revision.

After these first three chapters of introduction, in chapters 4, 5, and 6 we consider the technical issues needed to create ecologically plausible null hypotheses. Depending on the reader's interest, these may be the most daunting or the most important parts of what we have written. We begin this section with a prologue that introduces the key ideas. First, come the types of constraints we employ and why other constraints do not do the job. Second, we consider how to create the null matrices. The task is surprisingly difficult. Third, we consider the metrics—the particular measures we calculate to draw conclusions about the existence of patterns. In particular, we dismiss what we call "ensemble metrics"— measures that characterize an entire community with a single number. We find they typically hide interesting ecological patterns, even when they are there.

In the book's final section we return to the distributions of birds on islands, armed with the statistical machinery developed in chapters 4, 5, and 6. In chapter 7, we reconsider the birds of the islands of Vanuatu and the Galápagos. The former have few pairs of species belonging to the same genus—and it is within such pairs that we are most likely to find checkerboard patterns of mutual exclusion. We do not find them. In the Galápagos, there are many congeneric species sets. The patterns of their distribution are striking. Similar, congeneric species do not often co-occur. When they do, the sympatric populations evolve differences in beak size that likely reduce competition for seeds.

In chapter 8, we return to the Bismarck and Solomon islands, the locations that generated Diamond's ideas on assembly rules. The null models generate a list of species pairs where the number of observed co-occurrences is unusual, meaning that this (or smaller) occurs in fewer than 5% of the null models. As Diamond's critics had argued, interspecific competition is unlikely to generate some of these unusual pairs—they are ecologically and taxonomically very dissimilar. We show, however, that unusual pairs are disproportionately common in pairs that belong to the same genus—exactly the pattern one would predict.

Chapter 9 extends these ideas to species along an elevation gradient. The hypothesis is that along gradients, one species will replace another similar one at some perhaps arbitrary point. For example, one congener might occur below 1,000 m above sea level and the other only above 1,000 m. We present a case history to test this example. A second hypothesis is that there is some narrow band of, say, elevations where there is a general turnover in species. That is, several species may occur below 1,000 m and several only above 1,000 m. We test that idea too.

Finally, in chapter 10, we move from biogeographical patterns to those describing food webs. In particular, we consider the subset of a food web that describes which consumer species eat which resource species. For both island distributions and food webs, a widely studied metric is *nestedness*. Nested distributions occur when all the species that occur on an island also live on every island with a larger number of species. It does not hold if some of the species on the island with fewer species do not occur on one or more island with more species. Its introduction to ecology by Bruce Patterson and the late Wirt Atmar was in terms of nestedness being an obvious "signal" of ecological processes under which subtle—and so more interesting—patterns might hide.

For biogeographic distributions, those patterns include the checkerboard distributions. For food webs, they include *reciprocal specialization*. That is an awkward piece of jargon, we admit, but one made familiar in the morphological pairings of birds' beaks and the often strikingly shaped flowers into which they seem to fit perfectly. Co-evolution is the process normally posited to explain the pattern of reciprocal specialization. We show that such patterns are not common in nature.

CHAPTER TWO

Diamond's Assembly Rules

The working hypothesis is that, through diffuse competition, the component species of a community are selected, and coadjusted in their niches and abundances, so as to fit with each other and to resist invaders. — Diamond 1975

Here we examine Diamond's chapter in *Ecology and Evolution of Communities* (1975) in which he deduces a powerful, extensive role of interspecific competition in shaping how species occur in a group of islands. We first set the historical context. MacArthur had laid out a research program—in the sense of the philosopher of science Imre Lakatos (1978)—at whose core was an irrefutable idea that we could understand the large-scale patterns of nature and construct theories to explain them. Irrefutable ideas, of course, may be utterly trite and generate no testable ideas. This program, however, had great vitality in that it embraced a scientific approach to the study of ecological communities whereby theories could be tested by hypotheses that could be refuted or confirmed, spawning along the way large numbers of individually testable ideas.

Diamond's chapter in *Ecology and Evolution of Communities* was a long account of the distribution of birds on New Guinea and nearby islands, particularly the Bismarck Archipelago. In it, Diamond presented mostly simple natural history observations—which species occurred where, mostly—combined with fewer observations of which species had failed to colonize other places. What made the chapter so influential was its setting within the theoretical context of how competition shaped ecological communities. Most important of all was that Diamond distilled his observations into a small set of "rules" to describe the patterns of species co-occurrence. Out of complexity came simplicity.

Robert MacArthur, 1930–1972

Much of MacArthur's work is in his book *Geographical Ecology: Patterns in the Distribution of Species*, a volume that had a major influence on ecology. In his two books and a mere handful of papers, MacArthur laid out an approach to ecology of looking at large-scale patterns and explaining them in simple, mechanistic models. "Deriving patterns from simple differential equations," as he put it (Michael Rosenzweig, pers. comm.).

Colleagues and students attended MacArthur's memorial, organized in 1973 at Princeton University. Present were prominent and upcoming ecologists of the time, names who still exert a major influence today. Diamond, though just thirty-one years old, was among them. His presentation took up some 103 pages published in the memorial volume (Cody and Diamond 1975) and has been required reading for ecologists ever since.

In that volume, Diamond took up MacArthur's view with enthusiasm. On the basis of his observations of bird communities on the Bismarck Islands off the coast of New Guinea, Diamond proposed *community assembly rules* that, first, described what birds were found on particular islands and, second, provided a simple mechanism to explain them.

Special Islands and Their Birds

The main thrust of the volume's research was large-scale patterns. No paper illustrates this better than Diamond's. In his work on the birds of the Bismarck Archipelago near New Guinea, Diamond (1975) laid the foundation for rules that govern community composition. His detailed examination of the data convinced him that there were constraints on species composition, particularly on guilds, and he proposed a simple set of rules governing the assembly of communities.

Before we continue, we should answer two questions. Where are the islands, and what is special about them and their birds? Stretching from the equator, south to the Tropic of Capricorn, north and east of Australia, they are an unusual collection of tropical islands (see fig. 2.1). Only the islands to the west of New Guinea—those that inspired Wallace to write his "Sarawak law" paper—are as wet, tropical, and numerous. "Wet and tropical" means moist, humid forests, and forests that guarantee large numbers of species of birds (and many other taxa). "Numerous" means

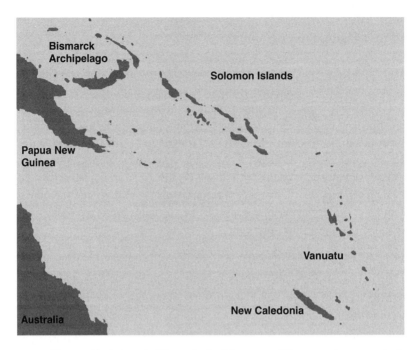

FIG. 2.1. Three archipelagos—the Bismarcks, the Solomons, and Vanuatu—play a central role in the ideas we present.

that there are many islands of broadly the same size and with broadly similar species. Nature herself casts species about these islands, providing us unintentional, uncontrolled, but nonetheless very interesting *natural* experiments on a huge geographical scale.

Three archipelagos play starring roles in our story. Just to the east of New Guinea lies the Bismarck Archipelago, whose largest island member is New Britain. To the southeast lie the Solomon Islands. The Vanuatu Archipelago lies still farther to the southeast.

The fruit doves provided the essential evidence for Diamond's ideas. New Guinea is home to 18 species of fruit doves, 12 of the genus *Ptilinopus* and 6 generally larger species of the genus *Ducula*. These species are all ecologically similar. They live in the canopy of trees and eat only soft fruit. At a single location in lowland New Guinea, Diamond could find as many as eight species. Interestingly, these eight species, whichever ones they were, formed a graded size series, with each species weighing about 1.5 times more than the next smaller species. Diamond gave one example of eight species whose masses were 49, 76, 123, 163, 245, 414, 592, and 802 grams (g). Apparently, members of the guild of fruit pigeons par-

P. insolitus   *P. viridis*

P. rivoli   *P. solomonensis*

P. superbus   *P. greyi*

FIG. 2.2. The species of *Ptilinopus* fruit dove.

titioned feeding stations on fruiting trees. (MacArthur had shown something broadly similar in his studies [1958] of the feeding habitats of warblers in eastern North American forests.) Smaller fruit pigeons could sit
on small limbs of the same tree and eat smaller fruit unavailable to the
larger fruit pigeons. Larger fruit pigeons consumed larger fruits unavailable to smaller members of the guild.

The other ten species replace various members of the lowland species in different habitats. That is, one or more species living in the New
Guinea lowland forest are replaced at higher elevations, or in different

habitats such as savanna, dry forest, or coastal forest. For instance, Diamond found that *Ptilinopus coronulatus* (75 g) replaces *P. pulchellus* (76 g) in high-rainfall areas, *P. iozonus* (112 g) replaces *P. superbus* (123 g) in open country, and *Ducula zoeae* (592 g) occurs at low elevations, whereas *D. chalconota* (613 g) occurs at high elevations. Any given weight range has the potential of being occupied by just one of from one to six members of the community, Diamond concluded.

Diamond found his most intriguing results on the islands off New Guinea (fig. 2.3). He counted 141 species of land birds on 142 islands in the Solomon Archipelago, including 19 species of doves and pigeons, 11 species of parrots, 6 species each of starlings, white-eyes, cuckooshrikes, and fantails. The Bismarck Archipelago held 150 species on 41 islands, equally rich in similar species. Indeed, these islands held 5 species of fruit doves in the genus *Ptilinopus*, a genus of 50 species that occur (or occurred, since one species is now extinct) in Southeast Asia, the Philippines, and Australia.

Figure 2.4 shows the distribution of the five species of *Ptilinopus* in the Bismarck and Solomon Archipelagos, plus a sixth species that occurs in Vanuatu, the islands to the southeast (fig. 2.2). We organize Diamond's records for the genus *Ptilinopus* in the Bismarck Archipelago in table 2.1.

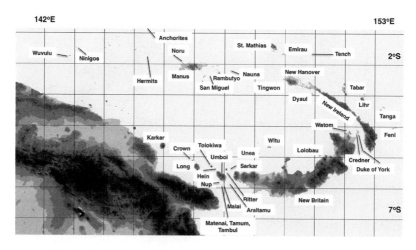

FIG. 2.3. The Bismarcks. For maps in this book we use the Mercator projection—a rectangular grid that does not represent equal areas correctly away from the equator. It works well for maps near the equator, however, which is where all our examples are located. Elevations are shaded in increasingly dark shades of gray at intervals of 100 m, 200 m, 500 m, 1,000 m, 1,500 m, 2,000 m, and 3,000 m.

TABLE 2.1 **Distribution of five species of *Ptilinopus* fruit doves on the Bismarck Islands (1 = present, 0 = absent)**

Island	Total of all species	Species/Occurrences P. superbus 12	P. insolitus 17	P. rivoli 9	P. solomonensis 22	P. viridis 3
New Britain	127	1	1	1	0	0
New Ireland	103	1	1	1	0	0
Umboi	83	1	1	1	1	0
New Hanover	75	1	1	1	1	0
Lihir	60	1	1	1	0	1
Tabar	60	1	1	1	0	0
Watom	58	0	1	0	1	0
Long	54	0	1	0	1	0
Lolobau	54	1	1	1	0	0
Manus	51	1	0	0	1	1
Dyaul	50	0	1	1	0	0
Duke of York	48	0	1	0	0	0
Tolokiwa	44	1	1	0	1	0
St. Matthias	43	0	1	0	1	0
Tanga	41	1	0	1	0	1
Feni	39	1	1	0	0	0
Sakar	36	0	1	0	1	0
Crown	33	0	1	0	1	0
Emirau	32	0	1	0	1	0
Witu	32	0	0	0	1	0
Rambutyo	30	1	0	0	1	0
Malai	21	0	0	0	1	0
Nauna	20	0	0	0	1	0
Unea	20	0	0	0	0	0
Wuvulu	17	0	0	0	1	0
San Miguel	16	0	0	0	1	0
Hermits	16	0	0	0	1	0
Ninigos	16	0	0	0	1	0
Credner	15	0	0	0	0	0
Tingwon	14	0	0	0	1	0
Tench	13	0	0	0	1	0
Noru	10	0	0	0	0	0
Anchorites	9	0	0	0	1	0
Tambiu	9	0	0	0	0	0
Midi	9	0	0	0	0	0
Nup	7	0	0	0	0	0
Tamum	5	0	0	0	0	0
Hein	5	0	0	0	0	0
Matenai	4	0	0	0	0	0
Araltamu	4	0	0	0	0	0
Ritter	4	0	0	0	1	0

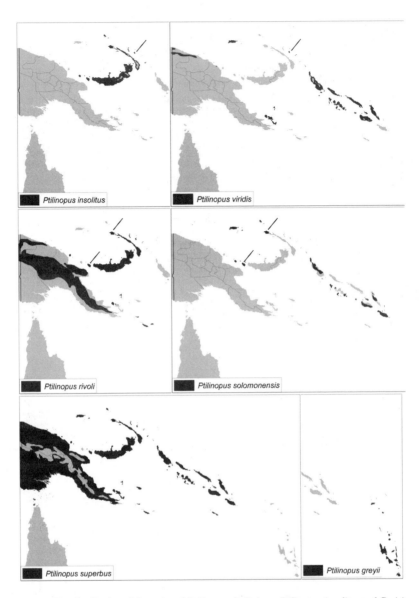

FIG. 2.4. The distribution of 6 species of *Ptilinopus* fruit dove. *Ptilinopus insolitus* and *P. viridis* co-occur on only one island (as shown by the line), despite the range of the latter entirely encompassing the former. Similarly, the range of *P. solomonensis* spans the range of *P. rivoli* within the Bismarcks and Solomons, but they only co-occur on two islands (also shown by lines). *Ptilinopus superbus* and *P. greyi* do not overlap in their ranges. This and all other maps combine data from several sources, including www.birdlife.org and www.hbw.com, but are adjusted where necessary by the data presented in Sanderson et al. (2009), who present the island-by-island presences and absences.

A 1 means the species was found on a given island, a 0 otherwise. We have also counted the total number of all land-dependent species Diamond found on each island. The numbers range from 4 to 127.

Consider *P. rivoli* and *P. solomonensis* in table 2.1 and figure 2.4. *P. solomonensis* has a range that spans over 3,000 km from the Schouten Islands off the north coast of West Papua (the part of the island of New Guinea that is Indonesia—and not shown in the figure), eastward through the Bismarck Archipelago, then southeast through the Solomon Archipelago. *P. rivoli* range has a similar span, from the Moluccas to the west of New Guinea, through New Guinea and to the islands to its southeast. The two species occur together on some of the Schouten Islands. Across all of the Bismarck and Solomon Archipelagos, they co-occur on just two islands—New Hanover and Umboi. Yet, *Ptilinopus* doves live on islands all across this archipelago, ranging from one island (Ritter) that houses the minimum number of species found (4) to New Britain, which has the maximum number of species (127).

Strikingly, the two ranges form a geographical "checkerboard" at two different geographical scales. More broadly, *P. solomonensis* is northwest and southeast of the large islands in the Bismarck Archipelago, while *P. rivoli* inserts its range in a central slice in between (fig. 2.4). At a finer scale, consider an example near the town of Rabaul at the eastern end of New Britain. That island has *P. rivoli*; Watom, 10 km offshore to the north, has *P. solomonensis*; whereas another island, Duke of York, 25 km to the east, only has another species, *P. insolitus* (which is also on Watom).

A second checkerboard involves the pair *Ptilinopus insolitus* and *P. viridis*. The former once again cuts a slice into the distribution of the latter, which occurs to the north and west, the south, and the east of it.

We cannot overemphasize the importance of a good map. As we finished this book, Connor et al. (2013) published a criticism of our work. They argued that, when ranges were considered, many of the checkerboards Diamond claimed were merely geographical separation. We agree in the case of *P. greyi*, the member of the genus that occurs in Vanuatu—islands to the southeast of the Solomons and Bismarcks. To be a checkerboard, species ranges must interlace—as do the black and white squares of a physical checkerboard. Connor et al. concluded that, for example, *P. rivoli* and *P. solomonensis* are just a simple consequence of the former having a more easterly distribution and the latter a more westerly distribution in the Bismarcks. They did not map any of the species they considered and analyzed the Solomons and Bismarcks separately. As figure 2.4

clearly shows, *P. solomonensis* occurs in the western Bismarcks and then in the Solomons, begging the obvious question of why it is missing from the larger islands in between.

The larger bodied doves in the genus *Ducula* are interesting too. Seven species occur only on New Guinea—*bicolor, chalconota, pinon, mullerii, rufigaster, spilorrhoa,* and *zoeae*—and are absent from the Bismarck and the Solomon Archipelagos. These islands have seven other species of their own that are absent from New Guinea (fig. 2.5). Six species are on large islands: *D. rubricera* is throughout the two archipelagos as is *pistrinaria,* while *finschii, melanochroa,* and *subflavescens* and are on the larger islands in the Bismarck Archipelago and *brenchleyi* on three islands in the southern Solomon Archipelago. So, these two sets of 7 and 6 *Ducula* species do not co-occur. A seventh species, *D. pacifica,* has a range that encompasses all of these 6 species yet barely co-occurs with any of these previous 6, generally being found on offshore islands off New Guinea and through the two archipelagos. It is on the large island of New Caledonia, widespread on the much larger islands of Vanuatu, and across the Pacific to the Fiji group—where again it is on small offshore islands, and eastward to the Cook Islands. On Vanuatu, an island group we discuss in detail in chapter 7, it overlaps with *D. bakeri* on the eight islands on which it lives. *D. bakeri* is generally restricted and more common at higher elevations than *D. pacifica.* By this time, you will likely predict that there is yet another *Ducula* on the larger islands of Fiji, and indeed there is—*D. latrans.*

What Is a Checkerboard Distribution?

The immediate answer to this question seems obvious: the pattern whereby two species occur widely but rarely co-occur. In the form of table 2.1, most islands have one species, few islands have neither species, and no islands have both species. Both species absent is less damaging to the claim of a checkerboard than both present. We readily understand why remote islands may be too far to be colonized or small ones be too small, or have the wrong habitats, to sustain the species.

There is something else, however, that one needs. If one species occurred only in the west and the other in the east, for example, that would not constitute a checkerboard, so a lack of islands with both species is not sufficient. The broad ranges have to overlap substantially. To show this overlap, maps of the species' distributions are essential. Diamond

D. spilorrhoa   *D. subflavacens*   *D. melanochroa*

D. brenchleyi   *D. finschii*   *D. rubricera*

D. pistrinaria   *D. pacifica*

FIG. 2.5. The species of *Ducula* fruit dove.

provided six maps with 22 species in his chapter for the island species we consider. Mayr and Diamond (2001) have 52 maps. In a recent critique of checkerboards, Connor et al. (2013) worried about range overlap but nonetheless produced not a single map for any of the species they considered. We will not repeat that mistake. Maps of all the world's bird species are readily available from Birdlife (http://www.birdlife.org/datazone) and from the IUCN Red List (http://www.iucnredlist.org). Both have excellent mapping interfaces and at resolutions to show even small islands with the archipelagos we discuss.

In addition to the *Ptilinopus* doves we have already mapped, consider the pair of *Ducula* pigeons shown in figure 2.6. *D. pacifica* and *D. pistrinaria* have ranges that span both the Bismarck and the Solomon Archipelagos. Draw a convex polygon around the former and it would entirely encompass the latter. Yet the two species co-occur on few islands in either archipelago—two islands in the Bismarck Archipelago and a cluster of small islands in the Solomon Archipelago.

Nothing beats a map to summarize the species distributions, their broad geographical overlap, and species co-occurrences. Nonetheless, this can get complicated for many species, so Diamond simplified things by introducing "incidence."

Incidence

We need more terminology. Diamond also discussed the *incidences* that described the occurrence of a species on islands in an archipelago in a simplified way. By *incidence*, we mean the range in the total numbers of species present on the islands where a given species occurs. Diamond noticed that some species occur only on islands with few species—typically small islands—while others occur only on species-rich islands—typically large ones. For instance, table 2.1 shows that *P. superbus* occurs on islands that hold from 30 to 127 species, whereas *P. solomonensis* spans the entire range—4 to 127. We return to incidences in more detail in the next chapter. For the present, some simple patterns are visible.

Diamond defined "high-S" species to be those species that occur only on species-rich islands. Twenty-seven species live only on either or both of the two largest islands, New Britain and New Ireland. Perhaps that is not surprising since these two large islands afford habitats—such as freshwater—that are not present on smaller islands. (As we have seen for *Ducula*, there are also species that occur only on the largest island of New Guinea itself.)

Rather more surprising are what Diamond called "supertramps." These species live only on islands with few other species. Within the Bismarck Archipelago, the dove *Ducula pacifica* occurs on five islands, none of which has more than 17 species. Earlier we discussed the patterns of this genus on islands with many more species.

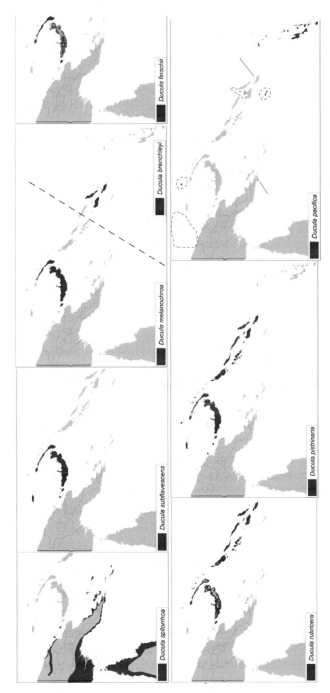

FIG. 2.6. The distributions of eight species of *Ducula. D. pacifica* has a huge range beyond that shown here and includes the Solomons, Bismarcks, and Vanuatu, where it is widespread. In the Solomons and Bismarcks, however, it is found only on small, scattered islands (lines and dotted lines indicate island clusters) and rarely co-occurs on islands with other congeners.

These patterns beg obvious questions. If one member of a genus could occur widely across many islands, then why not the others? And, when a species occurs so widely, why is it absent from many islands within that broad range?

Even communities on similar islands colonized from the same species pool sometimes differ greatly. The islands of Sakar and Tolokiwa are 46 km apart in the Bismarck Sea between New Guinea and New Britain (fig. 2.7). The islands differ in area by only 13 percent. Diamond reported that Sakar and Tolokiwa are similar geographically and floristically. The birds inhabiting each island came from the same pool of potential colonists. Tolokiwa, with 44 species of birds, lacks 3 of the 7 most abundant species found on Sakar (which has 36 species of birds). Sakar lacks 8 of the 15 most abundant species found on Tolokiwa. Just 25 bird species are common to both islands.

Moreover, a species found on both islands might occupy different habitats or occur in different abundances. For example, the fruit dove *P. insolitus* is present on both islands. On Sakar, Diamond wrote, *P. insolitus* is widespread. On Tolokiwa, it inhabits only midmontane forest. A related species, *P. solomonensis*, is also present on both islands and occupies similar habitats but is six times more abundant on Tolokiwa. For such similar islands, these are striking differences. How could such differences arise and be maintained?

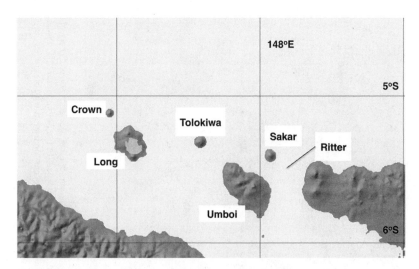

FIG. 2.7. The islands between New Guinea and New Britain. We shade these islands to indicate the volcanic cones on the islands and the large crater lake on Long.

FIG. 2.8. *Myzomela sclateri, M. pammelaena,* and *M. cruentata.*

Diamond reported that *P. solomonensis* was the sole member of its genus occupying Umboi Island in 1913, but sometime between then and 1933 both *P. insolitus* and *P. superbus* invaded the island. By 1972, the area and elevation occupied by *P. solomonensis* had contracted, and its continued existence on Umboi was uncertain.

There is another interesting pattern involving honeyeater species (fig. 2.8) on Long and its neighboring islands. First, consider the general patterns. Figure 2.9 shows the distribution of *Myzomela* species across the Bismarck Archipelago. Almost all the islands have just one species. *Myzomela pammelaena* is on 23 islands, *M. sclateri* is on 7 islands, and *M. cineracea* is on Umboi and New Britain. There are three classes of exceptions. First, as expected, some of the smallest islands have no species. Second, also as expected, the two largest islands have more species—two, in this case. *Myzomela pulchella* occurs inland at generally higher elevations than does the more coastal *M. cruentata* on New Ireland. On New Britain, *M. cineracea* occurs widely up to 1,200 m, whereas *M. cruentata* is now the more inland species.

The third class of exceptions is not expected. Three small islands—Crown, Long, and Tolokiwa—share *Myzomela pammelaena* and *M. sclateri*. Long's caldera collapsed in the mid-seventeenth century in one of the largest volcanic explosions of recent millennia. It likely exceeded the explosion of Krakatau in 1883 and ejected more than 30 km^3 of debris. The resulting crater lake covers 86 km^2 of the 328 km^2 island. Nothing would

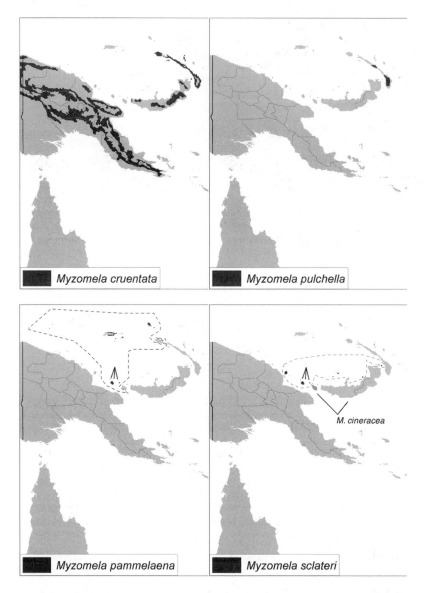

FIG. 2.9. The distribution of five *Myzomela* species in the Bismarcks. *M. cruentata* occurs on New Guinea and on New Britain, where it occurs in the highlands. On New Ireland, *M. pulchella* lives at higher elevations, while *M. cruentata* occurs in the lowlands. *M. pammelaena* and *M. sclateri* live on small offshore islands shown within the dashed lines. They co-occur only on Long, Crown, and Tolokiwa. *M. cineracea* occurs only on Umboi and New Britain, which lack the previous two species.

have survived. The present trees, mammals, and people of Long suggested recent overwater origins to Diamond. The birds of Long and the two nearest islands, Tolokiwa and Crown, also betray recent origin: Long has few of the expected large-island species, and all three islands have an excess of supertramps. Among the nine pairs of congeners on Long and its neighbors are two supertramp honeyeaters, with the larger, *M. pammelaena*, from the northern Bismarck Archipelago and the smaller, *M. sclateri*, from the southern Bismarck Archipelago. (Fig. 2.9 shows particularly well what Diamond means by *supertramp*—these two species occur only on small offshore islands, often close to, but never on, larger islands.) They constitute the sole pair of species that co-occur nowhere else, and they presumably first met on the Long group.

These are also the only two bird populations of the Long group that differ significantly in morphology from their relatives—surely their ancestors—elsewhere. Diamond *et al.* (1989) measured all available specimens of both species in museums, from the allopatric and putatively ancestral populations as well as from the sympatric populations of the Long group. The larger *M. pammelaena* is even larger on Long than in ancestral populations, whereas the smaller *M. sclateri* is even smaller. The weight ratio of the larger over the smaller is 1.52 on Long. Where the species do not overlap, the ratio is much lower—between 1.24 and 1.43. The two species co-occur abundantly in all habitats and at all altitudes of the Long group, and are often found in the same flowering tree.

The size shifts in these species on Long probably represent character displacement in response to each other's presence, putatively reducing the competition between them. Certainly, the shifts evolved in a very short period indeed—the approximately three centuries since Long's eruption.

The Theoretical Context

Diamond's natural history was not what triggered the fuss. Rather it was the relevance of his observations to the context of ecological theory at that time. In the late 1960s and early 1970s, the understanding of alternate, stable, invasion-resistant ecological communities of coadjusted species posed a major challenge in ecology. MacArthur's book *Geographical Ecology* explored the theoretical basis for such alternate communities. Diamond listed the relevant questions that he would answer in his chapter. In what follows, we deliberately use Diamond's terms and language whenever possible: "To what extent are the component species of a community mutually selected from a larger species pool so as to 'fit' with each other? Does the resulting community resist invasion? To what extent is the final species composition of a community uniquely specified by the properties of the physical environment, and to what extent does it depend on chance?" (Diamond 1975, 345). The last question is worth remembering. The "chance" part of it is aptly put. It would, however, ignite a vigorous debate for the next thirty years.

Diamond argued that assembly rules derived from species' competition for resources. Certain combinations of species maximized resource use and so prevented invasion by other species. He claimed that "communities are assembled through selection of colonists, adjustment of their abundances and compression of their niches so as to match the combined resource consumption curve of all colonists to the resource production curve of the island" (1975, 430). We have to say, in passing, that this claim is very ambitious and goes beyond what we will discuss. Diamond's working hypothesis was that "through diffuse competition, the component species of a community are selected, and co-adjusted in their niches and abundances, so as to fit with each other and to resist invaders" (430).

In what was to be a most controversial statement, Diamond wrote: "In a few instances, competition expresses itself in 'simple' checkerboard distributions, by which species replace each other one-for-one" (1975, 344). That is, species A and B form a checkerboard distribution when they do not co-occur on any islands, though there may be islands without either species. Diamond went on to list the assembly rules for *groups of related species* or *guilds*, a phrase that was destined to disappear from the vocabulary of his critics. In the great majority of species groups or guilds, competitive exclusion involves so-called diffuse competition, that is, the combined effects of several closely related species. Diamond's detailed examination of just four guilds revealed the following types of assembly rules for species communities:

1. If one considers all the combinations that can be formed from a group of related species, only certain ones of these combinations exist in nature.
2. These permissible combinations resist invaders that would transform them into a forbidden combination.
3. A combination that is stable on a large or species-rich island may be unstable on a small or species-poor island.
4. On a small or species-poor island, a combination may resist invaders that would be incorporated on a larger or more species-rich island.

5. Some pairs of species never coexist, either by themselves or as part of a larger combination.
6. Some pairs of species that form an unstable combination by themselves may form part of a stable larger combination.
7. Conversely, some combinations that are composed entirely of stable sub-combinations are themselves unstable. (Diamond 1975, 423)

Derived from the examination of just four guilds inhabiting a group of islands, these seven rules are complex and many think overly ambitious. We present the attack on these rules in the next chapter in some detail. Some issues, however, are obvious on first reading, even before we discuss the evidence for or against the rules.

Rule 1 is entirely empirical, derived as it is by looking at the lists of species that we record in our notebooks. So, too, is rule 5.

Rule 2, in striking contrast, is entirely about process and suggests a particular mechanism—competition—though it could be predation, competition, or indeed a very complex mix of the two. One could not demonstrate it from checklists. Rather, the rule requires an experiment or the good luck of having Nature provide one. (One such example involves Crowell's studies of islands off the coast of Maine. He demonstrated that, after he had released species of rodents on islands that naturally lacked them, the populations would go extinct eventually [Crowell and Pimm 1976]. Crowell's work was exceptional at the time in its geographical scale and experimental nature.) Diamond did have anecdotes for rule 2, as we relate below.

Rules 3, 6, and 7 introduce the undefined terms of *stable* and *unstable*, but we clearly sense their meaning. We routinely observe the former, but the latter only ephemerally. If we could do experiments, we might show that stable patterns would persist despite the introductions of other species, while unstable ones would not persist. In reviewing what *stability* means in ecology, Pimm (1991) discusses these and related patterns under the section of his book titled "Persistence."

Simply, the rules intermingle observed patterns with the mechanisms that produce them. If one takes the mechanisms as given, then it becomes clear that there are far fewer than seven rules. Moreover, the rules follow in an odd sequence. Rule 5 is the simplest two-species rule, while rules 3 and 4 are modifications of it. Rules 3 and 6 are essentially identical.

For all this, there are still several powerful ideas, and we state them purely in terms of what we might observe from a snapshot of species compositions. If one considers all the combinations that a group of related species can form, only certain ones of these combinations exist in nature. In particular, (1) some pairs of species never coexist; and (2) other pairs of species never coexist conditionally, that is, they do coexist but only when other certain species are present or when certain species are absent. Rule 1 is a *first-order* effect, that is, two species avoid each other, irrespective of which other species are present or absent. Rule 2 is a *second-order* effect: other species affect the outcome. One can imagine higher order effects too, when a yet different group of species modifies the second-order effects.

We betray our bias: the seven rules are far from perfect as stated, but they contain both a powerful description of potentially observed patterns and a compelling mechanism to explain them. In the next chapter, we encounter other scientists with much less charity.

Competition provides at least one of the possible underlying mechanisms, and as it does so, Diamond tells us he expects the rules to apply a fortiori to closely related species. Competition may often be an important mechanism, but it might not be the only one: it might be mutualism. The salient feature of the fictional clown fish Nemo is that he needs his sea anemones, so they occur together, not apart. (Art follows life.)

Predation could be the mechanism too. A predator might not survive without its prey, obviously. Less obviously, a prey might not survive without its predator, for in its absence another prey species, one that the predator prefers, might outcompete it. Indeed, two competing prey species might only survive if a predator is there to keep the dominant species in check. The familiar example of this is Paine's (1966) exclusion of the predatory starfish *Pisaster*. It is a common mechanism, but in an excellent review, Menge (1995) shows it is only one in a bestiary of complex patterns of how species indirectly affect each other. Nonetheless, much of Diamond's explanation for assembly rules had to do with competition. For the rest of this chapter, we continue to use Diamond's language and terms as well as his emphasis on competition. Diamond introduced a lot of jargon, much of it without definition, and much of it without any chance of our being able to measure or test empirically. In what follows, we put these terms in italics to alert the reader; should he or she not understand them or see how to measure them, then usually neither do we.

This example, and others from around the world, suggested to Diamond that species in a community are somehow filtered and their niches and abundances somehow adjusted, so that the community possesses some measure of stability, enabling the community to persist in time while resisting invaders. Diamond wrote: "It is then a logical extension of simple two-species distributional checkerboards to invoke 'diffuse competition'—i.e., the complex situations resulting from the sum of competitive effects from many other somewhat similar species. The power of this concept is that, in principle, it can explain anything. Its heuristic weakness is that, if it is important at all, its operation is likely to be so complicated that its existence becomes difficult to establish and impossible to refute" (1975, 348). We imagine his critics reading the words "difficult to establish and impossible to refute."

Diamond argued that to test the hypothesis of alternate, stable, invasion-resistant communities integrated by diffuse competition required a natural situation with several important properties. All were met by the island of New Guinea and its thousands of surrounding islands containing some 513 breeding land-dependent birds. Diamond wrote: "We shall see, that the distributions of most species can be neatly related to total species number in a community; and that, in a few cases, it is possible to relate species distributions to diffuse competitive effects from specific combinations of related species" (1975, 349). The key phrase "related species" meant groups of taxonomically related species—which are generally also ecologically similar or guilds—in which the species might not be taxonomically related, but which fed in similar ways.

Levins (1975) in the first chapter of the memorial volume wrote that the course of evolution of a species might not be comprehensible without reference to local community interactions. Reducing resources whether by a single species or by a "mutually adjusted guild of species, may be the principle mechanism by which invaders are competitively excluded from integrated communities" (Diamond 1975, 385).

Diamond's goal was to demonstrate that local species guilds were composed of coadjusted species that resisted invaders. In particular, he claimed that if the supertramps—those species able to exist on any island and whose dispersal abilities enabled them to reach all islands—were absent from a given island, then they must have been competitively excluded from that island by other species.

Though many authors chose subsequently to look at checkerboard patterns, Diamond actually considered checkerboards to be the least interesting: "To understand the role of competition in the assembly of communities, we shall begin with the most clearly illustrative but least important example, the distributional checkerboard. We shall then go through more complicated but still decipherable examples from four guilds that illustrate diffuse competition, the phenomenon of a species that is unable to fit into a community because it is excluded by specific *combinations* of other species" (1975, 387). Checkerboards occur when "two or more ecologically similar species have mutually exclusive but interdigitating distributions in an archipelago, each island supporting no more than one species."

Diamond found multiple examples and that the exceptions—where similar species coexisted—were often special circumstances. Examples were 2 species of cuckoo doves, *Macropygia mackinlayi* and *M. nigrirostris*, 2 flycatchers, *Pachycephala pectoralis* and *P. melanura*, the 2 species of fruit doves *Ptilinopus solomonensis* and *P. rivoli*, which we have already discussed, 5 species of small honeyeaters in the genus *Myzomela*, and 12 species of white-eyes in the genus *Zosterops*. The following example is one we can use to explore patterns in a quantitative way.

The Cuckoo Doves

Widespread from India to Australia, *Macropygia* cuckoo doves are arboreal, long-tailed, and they eat fruit. They live in the middle story of shaded

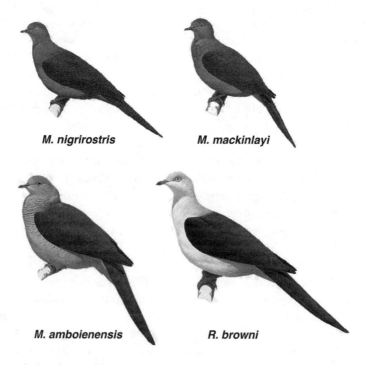

M. nigrirostris   M. mackinlayi

M. amboienensis   R. browni

FIG. 2.10. Cuckoo doves.

forests. *Macropygia mackinlayi* and *M. nigrirostris* are small—about 87 g, while *M. amboinensis* is medium-sized (150 g) (fig. 2.10). A related species, now put in the separate but closely related genus *Reinwardtoena browni*, is large (300 g). (In this book, we follow the taxonomic decisions of the *Handbook of the Birds of the World Alive*; www.hbw.com.) Diamond reported that the larger species can eat larger fruits, while smaller species can perch on small branches, enabling them to reach food unavailable to the larger species.

The broad-scale distributions show interesting patterns. *Macropygia mackinlayi* occurs from the western islands within the Bismarck Archipelago, then south and east through the Solomon Archipelago and the many islands of Vanuatu. Although it occurs on large islands in those last two archipelagos, it is absent from the two largest islands in the Bismarck Archipelago. *M. nigrirostris* occurs on those islands and also on New Guinea. *M. amboinensis* occurs from west of New Guinea, through the Bismarck Archipelago, but not the Solomon Archipelago, then south along the coast of Australia. *R. browni* is throughout the Bismarck Archipelago but nowhere else. As with the example of the two *Ptilinopus* doves, one species—*M. nigrirostris*—fills a surprisingly large hole in the distribution of another, very widely distributed species—*M. mackinlayi*.

As with the doves, the details within the Bismarck Archipelago are interesting, and table 2.2 shows their distribution there.

The cuckoo doves co-occur on some islands. Using Diamond's language, we see that Diamond derived "permitted combinations" and "forbidden combinations." For instance, he suggested that 2 of the 4 possible one-species combinations, 3 of the 6 possible two-species combinations, 2 of the 4 three-species combinations, and the single four-species combination are all forbidden combinations.

Of the islands with fewer than 30 species in total, only three have any cuckoo doves. *M. mackinlayi* and *M. amboinensis* are the only species that occur on their own. There are six possible combinations with two species. Three possibilities occur: *M. amboinensis* and *M. mackinlayi*, *M. amboinensis* and *R. browni*, and *R. browni* and *M. mackinlayi*; the other three possible pairs do not. Of the four possible trios that might occur, only two do so: *M. amboinensis*, *M. mackinlayi*, and *R. browni*; and *M. amboinensis*, *M. nigrirostris*, and *R. browni*. The four species do not occur together on any island.

Assembly rule 5 states that some pairs of species never coexist. In fact, *M. mackinlayi* and *M. nigrirostris* never co-occur. Note that these two species are of a similar size. Their distributions are shown in figure 2.11.

TABLE 2.2 *Macropygia* species (*amboinensis, nigrirostris, mackinlayi*) and *Reinwardtoena browni* in the Bismarck Islands (1 = present, 0 = absent)

Island	Total of all species	Species M. amboinensis	M. nigrirostris	M. mackinlayi	R. browni
New Britain	127	1	1	vagrant	1
New Ireland	103	1	1	vagrant	1
Umboi	83	1	0	1	1
New Hanover	75	1	1	0	1
Lihir	60	1	0	1	1
Tabar	60	1	1	0	1
Watum	58	1	vagrant	1	0
Lolobau	54	1	0	0	1
Long	54	1	0	1	0
Manus	51	1	0	1	1
Dyaul	50	1	0	0	1
Duke of York	48	1	1	0	1
Tolokiwa	44	1	0	1	0
St. Matthias	43	0	0	1	0
Tanga	41	1	0	0	0
Feni	39	1	0	0	0
Sakar	36	1	0	1	0
Crown	33	0	0	1	0
Emirau	32	0	0	1	0
Witu	32	0	0	1	0
Rambutyo	30	0	0	1	1
Malai	21	0	0	0	0
Nauna	20	0	0	1	1
Unea	20	0	0	0	0
Wuvulu	17	0	0	0	0
Hermits	16	0	0	0	0
Ninigos	16	0	0	0	0
San Miguel	16	0	0	1	0
Credner	15	0	0	0	0
Tingwon	14	0	0	0	0
Tench	13	0	0	1	0
Noru	10	0	0	0	0
Anchorites	9	0	0	0	0
Midi	9	0	0	0	0
Tambiu	9	0	0	0	0
Nup	7	0	0	0	0
Hein	5	0	0	0	0
Tamum	5	0	0	0	0
Araltamu	4	0	0	0	0
Matenai	4	0	0	0	0
Ritter	4	0	0	0	0

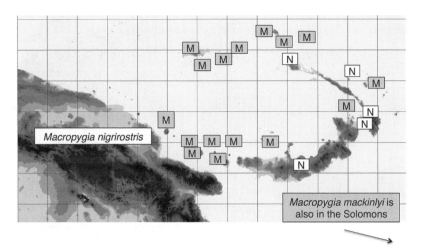

FIG. 2.11. The distribution of *M. mackinlayi* (M) and *M. nigrirostris* (N) in the Bismarcks is a perfect checkerboard. We omit vagrant records. A different subspecies of *M. mackinlayi* occurs extensively in the Solomons—the islands to the southeast of this map. The subspecies of *M. nigrirostris* found in New Guinea, *M. nigrirostris nigrirostris*, is different from the one found in the Bismarcks, *M. nigrirostris major*. We discuss the subspecies issues in chapter 8.

Assembly rule 2 says that "permissible combinations resist invaders." Diamond gave four examples of invasions that failed. Three species of cuckoo doves (*M. amboinensis, M. nigrirostris,* and *R. browni*) have been residents of New Britain throughout the last century. There is no resident population of *M. mackinlayi* on the island, however, vagrant individuals have been observed. *M. mackinlayi* is an abundant breeder on smaller islands within sight of New Britain and hence is likely to be a frequent invader. The combination *M. amboinensis, M. nigrirostris,* and *R. browni* thus appears to have resisted frequent invasions of *M. mackinlayi*. Other examples include failed invasions of *M. nigrirostris* on Karkar (occupied by *M. mackinlayi* and *R. browni)* and *M. nigrirostris* on Vuatom (occupied by *M. amboinensis* and *M. mackinlayi.)*

Patchy Distributions

Diamond also gave examples of conspicuous New Guinea birds that were present in certain disjoint areas but were absent from ecologically similar, well-explored areas elsewhere in New Guinea.

According to Diamond, "a skeptic can always dismiss these cases by suggesting that the distributional gaps are deficient in some unspecified but ecologically essential factor" (1975, 416). He argued that these distributional gaps were not due to any feature of the physical environment. These *patchy distributions* are distinctive features of tropical communities and are often not observed in temperate communities. He argued that competition for resources by members of the same guild could explain many of these patchy distributions. A slight advantage of one competitor over another, whatever that advantage might be, would ultimately lead to the absence of the weaker guild member. The addition of another guild member might cause that advantage, or perhaps a *constellation of competitors* in the community—in short, by diffuse competition.

Diamond explained the assembly rules for the fruit dove guild as (1) interspecific competition for resources, (2) *overexploitation strategies* whereby *permissible combinations* of species together consume resources and so thereby starve out competitors, (3) differences in dispersal rates of each species, and (4) the *unlikelihood of permissible combinations* once formed to be altered. Diamond's language was rich in the terminology of the day involving species' niches—and again, we employ his language deliberately, highlighting the jargon.

First, the niches of eight coexisting fruit doves and pigeons found in lowland New Guinea overlap. Second, Diamond noted that rarely does a species achieve its *potential niche* on a given island. The actual niche is always *considerably compressed*. The compression of various species' niches on the same island is correlated. Niche compression squeezes out other species and prevents invasions. Diamond recorded the size of the fruit consumed by each fruit pigeon.

From these observations, he suggested that communities assembled through selection of colonists whose abundances were adjusted and whose niches were compressed to match the production of the available resources. That is, there was room for only so many species within the same guild and no more.

How exactly are species that are similar in size to established residents excluded from a community? For instance, some supertramp species—those species that can reach the often-remote islands with few species on them—consume the same-sized fruit as some well-established residents. Why then do the supertramps not invade? Diamond suggested that supertramp species have relatively higher abundances than other species. This observation suggested that lowering of resource levels by members of a permissible combination could be an important mechanism of coadjust-ment and exclusion. *Overexploitation* is a viable strategy for a species faced with competitors, especially if the species regulates its population closely and has low reproductive potential but superior ability to survive scarcity. No supertramp could compete in such a community.

Competition for resources is not the only determinant of assembly rules. Between two competing species with mutually exclusive require-ments, Diamond said, equilibrium can be established on a species-poor island such that the fraction of time each species is present depends on population size and dispersal rates in addition to competitive abilities. Supertramp species have high populations and superior dispersal abili-ties and so can successfully occupy species-poor islands. Lastly, Diamond argued that, once assembled, permissible combinations resisted the re-placement of each species by other similar members of the guild. Thus, in his view, permissible combinations were stable.

The ease with which Diamond was able to construct assembly rules for guilds on islands was inversely proportional to the number of species in the guild, and directly proportional to the ecological isolation of the guild from other guilds. "If most of the competition faced by a particular spe-cies comes from just one or two other species, we stand a good chance of being able to construct assembly rules for the island communities. If in-stead there is diffuse competition from many species and no single com-petitor is overwhelmingly important, we remain baffled by patchiness, and we can make only statistical predictions of the probability that a species will occur on a particular island, from total species number and incidence functions" (Diamond 1975, 422).

Resource utilization was a major factor (Diamond 1975, 439) of the origin of assembly rules. Species in permissible combinations are *com-panions in starvation*, able to tolerate lowered resource levels. The ability to arrive on remote islands or those with few species and maintain large populations on them determines which species will occur there. Lastly, some combinations must be simply difficult to assemble, especially if some of the community members colonize infrequently.

Apparently, according to Diamond, by studying patterns of presence and absence on the islands in an archipelago, researchers could derive eco-logical assembly rules to explain patterns in communities on the islands, though some challenging unsolved problems remained. (That was quite an understatement!) Diamond suggested that the derivation of incidences needed more work, but he also discussed other problems. In his chapter, he used a heading, "Chance or predestination?" and with remarkable insight that suggested avenues of further investigation, he wrote: "At one extreme, the species composition of an island fauna might be uniquely determined by an island's physical properties. . . . At the other extreme, chance in the form of random historical events might play a large role in building up nonidentical communities that represent alternative stable equilibria. . . . Numerous findings suggest at least some role of chance" (1975, 440–41).

Summary

For the opening sentence of *Geographical Ecology*, MacArthur (1972) wrote: "To do science is to search for repeated patterns, not simply to accumulate facts." He died before the book appeared, but in a memorial volume, Diamond took up MacArthur's view with enthusiasm. On the basis of his observations of bird communities on the Bismarck Islands off the coast of New Guinea, first Diamond proposed *community assembly rules* that described what birds were found on particular islands. Second, he provided a simple mechanism to explain them.

The rules, as Diamond stated them, were logically rather tangled, combining observations and the mechanisms that might produce them. The discussion was also rich in the jargon of the day. Many of the specifics are surely overambitious, especially when it comes to "vacant niches" and "fitting species to resources."

They were also understated, in an important way. Although Diamond believed that competition was the key mechanism, there is nothing to prevent us from thinking about how predation or mutualism might produce large-scale patterns in nature.

All that said, however, at their simplest, there are two rules—and we still find them interesting. The first-order one is that a pair of species might occur together far less often or far more often than we expect. The obvious examples are the fruit doves, *P. solomonensis* and *P. rivoli*, and the two cuckoo doves, *M. nigrirostris* and *M. mackinlayi*, in the Bismarck Archipelago.

The second-order rule was that other species might make the first-order pattern conditional—that is, it would occur only in the presence or absence of some other species. A strong hint of this possibility comes from the observation that on large islands—such as New Guinea itself—more species of these genera occur in the same place. Adding more species is not likely to be the direct mechanism, of course. Rather, on larger islands with more opportunities to partition resources, species incompatible on small islands might find ways to coexist.

As MacArthur's research program hinted, the "repeated patterns" were across large geographical scales, ones where experiments would be impossible or highly exceptional. No one doubted that local-scale experiments—typically on the scale of a few square meters and for a few years—might find evidence for the postulated processes. All agreed that such experiments, however neatly crafted with statistical treatments and controls, were but proofs of concept. Yes, species might compete on a rocky shoreline or in a meadow, but that did not say anything about their geographical ecology.

Chance could play a role, too, meaning that there might be a large number of idiosyncratic explanations for particular presences or absences of species on islands. Diamond had followed MacArthur's research program, and it was not hard to imagine that there would be no end of students eager to elucidate the assembly rules governing all sorts of interesting, unique, and as-yet-unknown communities.

In 1979, however, Connor and Simberloff brought the party to an abrupt end. Assembly rules, they said, were either tautologies or trivial consequences of the original data. Diamond's assembly rules, they argued, did not exist; they were nothing more than figments of his imagination. Thus began an academic war fought in the public forum in the best ecological journals and books for more than thirty years.

As one of this chapter's reviewers put it—"Sorry to say, but reading this increased my sympathy to Conner and Simberloff's reaction. Some things in science are meant to do their work and disappear into the evolving fabric of our knowledge." We understand the point completely. As we wrote about these ideas forty years after their publication, we became uncomfortable too. Above, we have italicized some of the words we find difficult to define or to test easily with observations or experiments. Dissecting our own work from the mid-1970s, readers would surely find similar inadequacies, of course. The point of hindsight is not to berate authors for not clearly seeing the advances of future decades. Rather, the purpose is to gain foresight in how best to construct present work to stand the tests of time. And, of course, to understand how ideas develop, are tested, and generate debate and synthesis.

Above all is the overarching ambition of these ideas. As we explained at the start of the previous chapter, asking which species occur where and why are fundamental questions. That competition molds the composition of ecological communities at large geographical scales is a bold, exciting idea. It is certainly not the only explanation for species occurrences, but it is possibly an important one. In much the same way that evolution is ubiquitous, but Darwin and Wallace needed the simplicity of island systems to see its consequences clearly, then these extraordinary numerous and species-rich islands may permit our elucidation of the competitive process more generally. That is why the debates that follow merit the attention we now give them.

CHAPTER THREE

The Response of Connor
and Simberloff

We challenge Diamond's (1975) idea that island species distributions are determined predominantly by competition as canonized by his "assembly rules." We show that every assembly rule is either tautological, trivial, or a pattern expected were species distributed at random. — Connor and Simberloff 1979

N o ecologist reading the first two sentences of the abstract of Connor and Simberloff's paper (as quoted in the epigraph above) could set the article aside for bedtime reading. This was not just an assault on Diamond's work but on community ecology itself. After all, if communities were not in some way organized—or if we could not tell the difference between organization and disorder—then why would anyone want to study ecological communities?

Later, Simberloff was to feel himself "viewed as a professional crank" (Quammen 1996, 428), but that is unnecessarily self-deprecating. Some ecologists were finding patterns everywhere, although some were simply consequences of random processes, and Simberloff had an energetic, rigorous research program for sorting the wheat from the chaff, much of which has been uncontroversial.

Connor and Simberloff submitted their rebuttal in June 1978 to the journal *Ecology*, and the paper appeared in December 1979.

The backlash against Diamond had several dimensions. The first was a criticism of the way Diamond presented the assembly rules. It echoes comments we made in the previous chapter, but far more stridently.

Second, Connor and Simberloff calculated the chance that the two species of similarly sized cuckoo doves of the last chapter would co-occur on any of the islands. We will show there is about a 1 in 40 chance—smaller than the conventional 1 in 20 that is the standard benchmark for claiming statistical significance. What could be wrong with the result?

Third, a major separation between the two camps involved the role of prior knowledge about the species concerned. In the last chapter, we showed that Diamond used details about body weights and behavior to single out certain pairs of species for further analysis. Pick up almost any field guide, find the plates that show very similar species and your heart sinks: how will I tell them apart? Looking down the list of species on the Bismarck Archipelago, for example, the two cuckoo doves would immediately stand out as one of a mere handful of species very difficult to separate in the field. "If you are going to look for checkerboard distribution, this handful of species pairs is where to look!" shouts experience. Connor and Simberloff dismissed such arguments.

Fourth, Connor and Simberloff demanded that the possible occurrences of all species, species pairs, triplets, and so on be examined, irrespective of any prior expectations of what might or might not co-occur. They concluded that by chance alone we would not observe many possible combinations of species. In drawing this conclusion, they created models of how to construct incidence matrices under various ecologically reasonable constraints, but that is a topic for the next chapter.

The Backlash

Connor and Simberloff launched their attack by singling out assembly rules 1 and 5, which they labeled (a) and (e), respectively. We repeat them here using the notation employed by Connor and Simberloff (1979, 1132):

> Rule (a). If one considers all the combinations that can be formed from a group of related species, only certain ones of these combinations exist in nature. . . .
> Rule (e). Some pairs of species never coexist, either by themselves or as part of a larger combination.

Connor and Simberloff wrote: "These rules are identical, except that rule (a) is restricted to related species and rule (e), though unrestricted taxonomically, is concerned only with pairs of species" (1979, 1133). In rule (e), Diamond (1975), by deliberately not restricting it to "related" pairs, created an opening Connor and Simberloff exploited. Moreover, later investigators also created metrics based on an analysis of all possible pairs of species, not just those related taxonomically or in the same guild.

How Likely Are Checkerboards?

Next, Connor and Simberloff argued that the absence of some species pairs was indeed trivial. The 41 Bismarck Islands have 141 species of birds and so $(141 \times 140)/2 = 9{,}870$ possible pairs. Some pairs of birds never co-occur on any islands, because one would need 9,870 islands to represent each pattern just once. Thus, rule 5 is a trivial observation, one we are likely to find in any real situation, and it requires no special mechanism to generate it. That said, Diamond was clearly writing about closely related species—those in the same genus, in fact. There are far fewer pairs of congeneric species, a point to which we return later.

Diamond used five examples to demonstrate rule 5. One, doves of the genus *Macropygia*, consisted of two species whose distributions formed a checkerboard, that is, neither occupied the same islands. *Macropygia mackinlayi* lived on 15 islands, *M. nigrirostris* on 5 islands, and neither occurred on 21 islands. Connor and Simberloff's next criticism came as they computed the probability of the species not co-occurring. First, count the possible ways of placing 15 populations of *M. mackinlayi* on 31 islands. We chose 31, not 41, because there are 31 islands with more than 10 species on them. Only such islands have cuckoo doves (and not all do). Islands with 10 or fewer species do not.

We calculate this by noticing that there are 31 islands on which to place the first population of species *M. mackinlayi*. Once the first population of *M. mackinlayi* has been placed, there are 30 islands left without *M. mackinlayi* and so 30 ways of placing the second population. If we continue in this way from 31, 30, 29, down to 17, then we will have placed the species on 15 islands. Sixteen islands will remain without *M. mackinlayi*. In mathematical notation, the number $K! = K \times (K - 1) \times (K - 2) \ldots 3 \times 2 \times 1$. So 31!/16! is all the ways of picking 31, 30, 29, . . . , 3, 2, 1, islands, divided by the ways of picking the 16, 15, 14, . . . , 3, 2, 1 islands that we do not care about because we only need the first 15 islands.

We do not care in which order we pick the islands. That is, it does not matter whether the islands are picked to house species *M. mackinlayi* in sequence as 1, 2, 3, 4, 5, 6, 7, 8, 9, 10, 11, 12, 13, 14, 15, or 1, 2, 11, 12, 13, 14, 15, 3, 4, 5, 6, 7, 8, 9, 10, or any one of the vast number of possibilities—15! = 1.307×10^{12}. Actually, it only matters that these islands have *M. mackinlayi*.

So, the number of ways of arranging *M. mackinlayi* on these islands is 31!/(16! 15!) or 300,540,195 possibilities. The same logic gives us 31!/

(26!5!) = 169,911 possible ways of picking the islands that house species *M. nigrirostris*.

The product of these two numbers, about 10^{10}—a mind-boggling large number—is the number of possible ways to place two species. Some of those possibilities will have both species present on one or more islands. In fact, we do not need this huge number. What matters is how many of the 169,911 ways in which we pick the islands that have species *M. nigrirostris* on them but do not have *M. mackinlayi* on them. *M. mackinlayi* is absent from 16 islands. There are 16 ways of picking the first absence, 15 the second, down to 12 to pick the fifth (and final one). Again, the order of the absences does not matter. So there are 16!/(11!5!) = 4,368 ways.

Therefore, the probability that the two species do not co-occur is 4,368/169,911 or about 0.026. The checkerboard distribution could have arisen by chance alone about 1 in 39 times.

Suppose, for a moment, that we had done the previous calculations on all 41 islands. The same recipe would now show that the checkerboard distribution could have arisen by chance alone about 1 in 11 times. This should not surprise you. Suppose we had included every island in the Bismarck Archipelago—likely hundreds or thousands of them, most very tiny. The chance that two species would not meet across this large array of islands would be very high. For the benefit of future discussions, what this means is that the conclusions we draw about patterns depend on the assumptions we make about species incidences—the range of islands over which they occur.

Prior Expectations

A major controversy ensued. This 1 in 39 is generally considered to be statistically significant because it is less than 1 in 20 (a p-value of 0.05). Diamond argued that he had selected two doves, *M. mackinlayi* and *M. nigrirostris*, because of their similar sizes, feeding habits, and abilities to disperse broadly, and because they did not occur on islands that hold ten or fewer species in total. That is, the choice was made a priori.

Connor and Simberloff would have none of this. Their colleague Don Strong made the point succinctly: "if you pull the handle of a slot machine enough times, you shouldn't be surprised that eventually three cherries will pop up" (Strong, pers. comm.). Connor and Simberloff noticed that with 9,870 possible species pairs, finding one pair that should co-occur just once in 39 tries is hardly unusual. There should be more than 250 of them. Most of these would be ecologically meaningless, for they would include completely unrelated species that did not co-occur for varied reasons, but certainly not because of the presence of the other. For instance, the bittern *Ixobrychus flavicollis* and the white-eye *Zosterops griseotinctus* do not co-occur on islands in the Bismarck Archipelago, even though they span overlapping ranges (islands with 16 to 127, and 20 to 54 islands, respectively), but they are very different in their habitats and feeding methods. No one was about to ascribe their checkerboard distributions to interspecific competition. Clearly, what mattered were the prior ecological assumptions and knowledge one brought to the analyses.

Whatever the merits of the two viewpoints, the confrontation was an ugly one. Diamond had implied that he had picked his examples a priori using preexisting natural history and had not cherry-picked them, but we have no independent proof—just his word "as a gentleman" and our individual birding experiences. Intimating that he might have looked at hundreds of possibilities and picked just the ones that looked good was a personal insult, even if unintended.

Does this dustup seem too arcane to matter? Scientists use prior expectations routinely. Think of statistical tests that can be either two-sided (Is treatment A different from control B?) or one-sided (Is treatment A to be larger than control B?). At some predetermined level, often 1 in 20, a test might be significant as a one-tailed test but not as a two-tailed one. The simple point is that the way we frame our expectations of the data we analyze affects the conclusions we draw from them.

So, to give equal time to both sides, what follows if we accept that we do not know where to look for checkerboards, that we have no more expectation of finding them among cuckoo doves of roughly the same body size as among any other pair of species? To answer that question, Connor and Simberloff turned to complete lists of species on islands, not just what they considered to be cherry-picked examples.

The Analysis of Vanuatu

When Connor and Simberloff requested the data, Diamond denied them access on the grounds that he was about to publish them. Connor and Simberloff (1979) wrote: "We regret we cannot use the same Bismarck data which Diamond first used, but its publication has been delayed by various unforeseen complications (J. M. Diamond, pers. comm.; E. Mayr, pers. comm.)." Those data became available as machine-readable matrices when, with Diamond, we published them in 2009—that is, thirty years later (Sanderson et al. 2009). The unavailability of the data clearly contributed to the increasing hostility. In the end, Connor and Simberloff used three different data sets. Among these data were the birds of Vanuatu, an archipelago for which there are survey data on 28 islands containing, in various combinations, 56 land birds (Diamond and Marshall 1976) (see fig. 3.1).

The observed presence-absence matrix is approximately half full. Unlike the islands off New Guinea, no particular checkerboard patterns stood out. For 56 species of land birds, there are $56 \times 55/2 = 1,540$ possible pairs of birds. Connor and Simberloff asked how often each pair co-occurs across all 28 islands. This metric considers all possible pairs of birds, not just those in guilds, or in the same family or genus, and lumps them all into a single sum. A priori ecological similarity is discarded.

Because there are 28 islands, it is possible that some pairs of birds could never co-occur, co-occur once-and-only-once, twice-and-only-twice, three-and-only-three times, and so on up to 28-and-only-28 times. Thus, there are 29 possible ways each pair of birds can co-occur—including not co-occurring at all and so forming a checkerboard. Connor and Simberloff computed these 29 numbers from the observed incidence matrix and then compared the observed 29 numbers to the average found in the sample null space of 10 nulls. We discuss how they produced those nulls presently.

Like us, you may be thinking: 10 nulls? Why so few? Later, we generate them by the millions. That was a practical limit of computing power of the time.

In the observed incidence matrix, 63 pairs of birds never shared any island—they formed checkerboards. Consider the brown goshawk (*Accipiter fasciatus*) that occurs only on the most southern island, Aneityum, with 31 other species. Because there are 56 species, we know that the brown goshawk forms 24 checkerboards—one with each of the remaining 24 species $(56 - 31)$ that do not live on this island. None of those 24 species are in the same genus, and the only other two birds of prey, the swamp harrier (*Circus approximans*) and peregrine falcon (*Falco peregrinus*), feed in very different ways.

Similarly, the Santa Cruz ground dove (*Gallicolumba sanctaecrucis*), the Santo mountain starling (*Aplonis santovestris*), and the thicket warbler (*Cichlornis whitneyi*) occur only on the island with the most species,

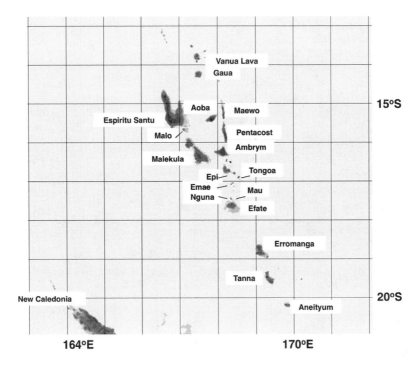

FIG. 3.1. The islands of Vanuatu. We label only those with 30 or more species of birds.

Espiritu Santo, occupied by a total of 50. Therefore, these 3 species do not co-occur with 6 species, thus giving rise to another 3 × 6 = 18 checkerboards. Just from these four single-island endemics, we identify 42 checkerboards.

There were 212 pairs of birds that co-occurred once-and-only-once, a smaller number that occurred twice, and so on. One of those pairs that only co-occur on one island is the parrotfinch *Erythrura trichroa* and the cuckooshrike *Coracina caledonica* whose distributions in Vanuatu appear in figure 3.2. The former has a large range that stretches from Australia to Japan and includes Papua New Guinea, New Caledonia, and the Solomon Archipelago. The cuckooshrike also occurs in New Caledonia and the Solomon Archipelago. That they co-occur on only one island (Erromanga) may seem surprising given their wide ranges both in Vanuatu and elsewhere, but no one suggests this may be due to one species' effect upon the other. They are not closely related, and differ in size and natural history: the finch feeds near the ground on grass seeds; the cuckooshrike is insec-

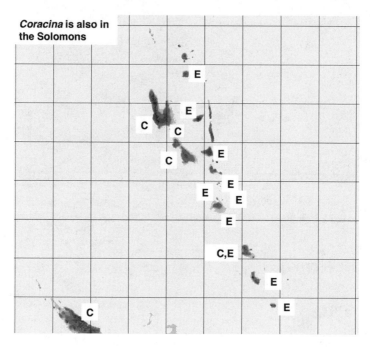

FIG. 3.2. The distributions of the parrotfinch *Erythrura trichroa* (E) and the cuckooshrike *Coracina caledonica* (C) in Vanuatu. This is a near checkerboard but surely an ecologically meaningless one.

tivorous and feeds in the canopy. *E. trichroa* occurs on three islands in the Solomon Archipelago, and *C. caledonica* is present on each one of them.

In fact, there is nothing unusual about 63 species pairs not co-occurring, nor is it unusual that there are even more possibilities for species to overlap on just one island. Simply, one expects there to be many checkerboards by chance and for very few of them to involve ecologically meaningful pairings.

What happens if one looks for checkerboards where the pairings might be ecologically interesting? Connor and Simberloff repeated the numbers-of-pairs but restricted their attention to birds in the same family. There are 99 pairs of species that are in the same family in one or another of the 15 families present. Once again, there was no difference between the numbers of pairs in the observed incidence matrix and that derived from the sample null space. Connor and Simberloff wrote: "In a nutshell, there is nothing about the absence of certain species pairs or trios, related or not, in the New Hebrides [now Vanuatu] that would not be expected were the birds randomly distributed over the islands as described above. Since there are so many possible sets of species, it is to be expected that a few sets are not found on any islands; this does not imply that such sets are actively forbidden by any deterministic forces" (1979, 1134).

There are nine genera that contain two species and one genus that contains three species. The species involved account for over a third of the species on these islands. It is within these species that we would expect to find checkerboard distributions, but Connor and Simberloff did not comment on them. In writing this book and having become familiar with the species involved, we find this omission curious. Not only are there many congeneric pairs, but they involve some of the same genera that form checkerboards in the Bismarck and Solomon Archipelagos. These are exactly the species for which one should expect there to be checkerboards.

The observed patterns, however, are almost completely the opposite of that expectation as one could imagine. There are two duck species of the genus *Anas*. Their ranges overlap completely, probably because of the presence of fresh water on certain islands. There are two *Ptilinopus* fruit doves—the same genus that forms the checkerboards described in the previous chapter for the Solomon Archipelago. One, *P. greyi*, occurs on every one of the 28 islands; the other, *P. tannensis*, on 23 of them. Two species of the large doves in the genus *Ducula* (which we discussed in the previous chapter), two *Halcyon* kingfishers, two *Lalage* trillers, two *Rhipidura* fantails, and two *Aplonis* starlings also overlap completely. Two species of white-eyes, *Zosterops*, occur on 26 islands each and so must overlap extensively, whereas two species of parrotfinch, *Erythrura*, occur on 10 and 14 islands, respectively, and are found on six of the same islands. Of the three species of *Collocalia* swiftlets, one is found on 26 islands, a second on 20, and the third on 11, and again there is substantial overlap.

Connor and Simberloff carefully detailed two more examples: West Indies birds (211 species on 19 islands; data from Bond 1971) and West Indies bats (59 species on 25 islands; data from Baker and Genoways 1978). The graphs in their paper told the story—there was absolutely no significant difference between the numbers of pairs in the observed incidence matrix and the numbers of pairs derived from the sample null space. A look at the two additional figures left no doubt.

The inescapable conclusion was that there was no support for Diamond's (1975) assembly rules 1 and 5 (rules *a* and *e*, respectively, in their analysis). All three examples fit the "random hypothesis." Connor and Simberloff were not suggesting the distribution was, in fact, random. What they showed was that there was no significant difference between the observed incidence matrix and one chosen randomly using the three constraints. There might indeed be a mechanism or mechanisms acting to structure the community, but statistically there was no way of discerning whether the observed patterns differed from chance expectations. The results presented by Connor and Simberloff had widespread and long-lasting implications for community ecology.

Connor and Simberloff next took on assembly rule 2: Permissible combinations resist invaders that would transform them into forbidden combinations. This rule is clearly a deduction from the definitions of "permissible combination" and "forbidden combination," plus Diamond's explanation of the assembly rules. Permissible combinations are those that exist in nature, and forbidden ones do not exist.

Connor and Simberloff wrote: "Nor is there compelling experimental evidence that any particular combination is actively forbidden by any force(s), or actively 'resists' transformation." They continued: "In no case is there evidence that active resistance occurs, unless one begins with the assumption that a distributional gap must be explained by active resistance, in which instance we have a tautology and why bother with the exercise of producing evidence?" (1979, 1136). Having built up momentum by using the null model complete with statistical analyses, they similarly and more quickly demolished the remaining assembly rules.

Summary

Connor and Simberloff were bitterly critical of the way Diamond had formulated his assembly rules, but they went further, arguing that the key observation—that species pairs form checkerboards—was trivial. In the final section of their paper, titled "Coda," they concluded their arguments: "Clearly, the assembly rules do not compel us to posit interspecific competition as a major organizing force for avian communities. That such an all-encompassing theory should be built on so little evidence invites an examination of the procedures used in its construction, and one point stands out. At no time was a parsimonious null hypothesis framed and tested. . . . Diamond (1975) assumed competition to be the primary determinant and then sought post facto to rationalize the observed data in the light of this assumption" (1979, 1138).

Connor and Simberloff took the example of the checkerboard distribution of the two cuckoo doves in the Bismarck Archipelago head on.

Yes, the chance that the two species did not occur on any islands was unusual—about 1 in 39—but they dismissed this. At issue is whether this example was selected, a posteriori, because it appeared special or whether there were grounds for picking it a priori.

If we have n species, there are $n \times (n - 1)/2$ possible pairs—45 with just 10 species. For any specific example—Vanuatu, for example—a fraction of these pairs will likely involve pairs that do not co-occur, or perhaps co-occur on one or a few islands. Most will involve species that have no ecological relationships to each other. Checkerboards, in themselves, are not noteworthy. What makes them special is when they appear in unusual numbers between species that are ecologically similar in some way.

The previous sentence may seem innocuous—but it sets the seed for bitter controversy and technically difficult discussion. First is the issue of being "ecologically similar."

Stuart went bird-watching in the Bismarck Archipelago for the first time in 2010. Looking at the list of species that occur there, he salivated over the new species to be added to his life list: the cuckoo doves and fruit pigeons—and other species Diamond discusses—jump out. "Cuckoo doves are brown with long tails and are nearly the same size—how do I tell them apart?" Figure 2.11 replies: one does not have to; they do not co-occur! It is an answer familiar to birders worldwide who worry about how to separate similar species. In practice, they often do not occur in the same places or at the same elevations, and if they do, their habitats are different. To answer Don Strong's quip about the slot machine lever, it is as if one pulls it just once and three cherries pop up on the first try: the cuckoo doves are exactly the kind of species pair one might expect to find on a checkerboard distribution. Bird field guides have maps next to the plates and descriptions of habitats precisely because this information is often so useful.

Importantly, such insights do not always work; the various congeneric pairs on Vanuatu confirm that. So how can we put these insights into a scientific framework that compels those whose prior expectations differ?

Second, stating that something happens unusually often begs our stating what our expectation is. That would require some statistical model of what should happen—and this topic is the subject of our next chapter. How exactly does one create a null model for species occurrences on islands?

PART II

A Technical Interlude

In the previous chapters, we introduced the idea that the patterns of presence and absence of species on islands might provide evidence of the power of competition shaping the co-existence of species—and do so on large geographical scales. Armed with the necessary statistical machinery, we then hoped to tackle some other issues of large-scale, complex community organization. Our approach in the earlier chapters was to introduce the problem in simple, intuitive terms. To further expand our ideas, considerable technical details are required. The next three chapters provide them. With those details in hand, we return to consider new analyzes of assembly rules for communities and expand the lessons learned to other topics involving ecological pattern. The point of this preamble is to permit the reader to skip the technical details of part II, if need be.

Our objective is to decide if the distribution of a set of species across a collection of islands is special in any way. To do this, we must decide what distributions might occur if we randomly placed species on the islands. This is a two-step process. First, we must create a random collection of communities. This involves the formation of random incidence matrices, using those "sensible constraints" we mentioned previously. Chapter 4 explores the constraints, and chapter 5 considers the difficulties in creating random incidence matrices that are subject to these constraints.

The second step requires that we determine a metric and a statistic that measures the discrepancy between the observed value of that metric and its distribution as generated by chance. Chapter 6 addresses that subject.

We must complete both steps adequately: a large number of random samples will not overcome the fundamental flaw of sampling the full null space incompletely or nonrandomly. Similarly, having the complete collection of every possible random community will not overcome a poorly chosen metric. The analysis of pattern in species co-occurrences therefore requires an understanding of binary matrices, that is, two-dimensional matrices filled with 0's and 1's representing absences and presences, respectively. Moreover, understanding how these matrices work is essential to unravel and appreciate the research efforts and results obtained to date.

Chapter 4 considers the constraints we might impose on a random matrix. The simplest way is to assume only that there is some fixed number of interactions or some fixed probability of there being a species on an island. For example, there are 888 presences of 56 species on 28 islands in Vanuatu—in other words, it is about half full. (To be full would require 56×28, or 1,568 presences.) One could take these 888 species-island presences and scatter them about each species-island matrix at random. Alternatively, one might chose each species on each island to be present with a probability of about 0.57 (= $888/(56 \times 38)$). The latter method would give an average occupancy of 888 species-island presences across a large number of random matrices, but individual matrices would differ. The former method produces 888 species-island presences every time, of course.

No one has applied such a simple model to the species-on-islands problem, but these models do appear in studies of food webs. Broadly, the former—keeping the number of interactions constant—is what one of us called his null model (Pimm 1982), while the latter method is what Cohen called his Cascade model (Cohen and Newman 1985). We return to food web patterns in chapter 10.

An additional level of constraint involves the number of species on islands and the number of islands occupied by each species. Thus, in the randomly generated incidence matrices, we assume that if a species actually occurs on k islands, it must also occur on k islands in each of the randomly constructed matrices. Moreover, if an island actually supports j species, so it must in all the randomly constructed matrices. In chapter 4, we present the arguments of those who think one such constraint is sufficient, but we consider both are necessary. Simply, every random matrix should have the same totals of species on islands and islands with species as the observed incidence matrix.

A more controversial constraint involves that of species incidence. In the example of cuckoo doves (in chap. 2), we noticed that no species of cuckoo doves occurred on islands with very few species on them. We then calculated the probability of two species occurring on as few islands as observed—none in this example. Although we were only looking at two species, we noticed that the assumptions we made about a species' incidence made a considerable difference as to whether we thought this perfect checkerboard was unusual or not. In the simple calculation we provided, a perfect checkerboard occurred as often as 1 in 11 if we included all 41 islands, but only 1 in 39 if we restricted the analysis to the 31 islands large enough to hold cuckoo doves.

Should we further constrain the random incidence matrices to have the same species incidences as those observed? Our answer is no for two reasons.

First, allowing species to occur on islands with few species produces conservative results—ones less likely to conclude that an observed checkerboard is unusual. (The example of the cuckoo doves illustrates this.) The second reason has to do with the supertramps—those species that occur only on islands with few species. Suppose we were to prevent their occurrences on islands with many species in the random matrices. Here, we would exclude the very result we find interesting—the likely competitive exclusion of these supertramp species by similar species that occur on the species-rich islands. This would be an example of what Colwell and Winkler (1984) called the Narcissus effect.

In Greek mythology, Narcissus was a beautiful young man who spurned his admirers, then fell in love with his own reflection in a pool and came to a bad end. Colwell and Winkler were not cautioning against graduate students who spend too much time in front of mirrors and their propensity to come to bad ends. Rather, they warned against the dangers of using species pools (clever pun) and the set of species that could occur on islands that already reflected (another pun) the influence of competitive exclusion. For both of these reasons, we do not use constraints on species incidences in creating random incidence matrices.

Chapter 5 considers how to fill in the presences and absences in constrained random incidence matrices. Connor and Simberloff used a mere ten random matrices, in part a consequence of the limitations of computer power back then. So perhaps one might now compute the entire set of possible constrained matrices. One cannot, however—there are simply too many of them—so instead one has to sample those possibilities. Two kinds of algorithms do so.

Connor and Simberloff used a *swapping* method that started with the observed incidence matrix and then swapped pairs of entries to retain each row and column sum. When only a few swaps are applied, the random matrices are very similar to the observed one. The conclusion that the observed patterns are not different from chance is simply inevitable—

and so is uninteresting. Chapter 5 introduces *construction* methods that seek to produce large numbers of random incidence matrices that are representative of the vast, but unknown, range of possibilities. What is required, however, is a method that produces independent random incidence matrices that sample the entire collection of all such matrices uniformly. Fortunately, such a method is available, and we describe it in chapter 5. We use this method in all subsequent chapters.

Chapter 6 reviews the history of the different metrics, or measures, that describe the observed and random incidence matrices. The history has been to describe communities with *ensemble* metrics—those that describe the entire statistical distribution of, for example, the number of pairs of species that form perfect checkerboards, co-occur on just one island, on two islands, up to co-occurring on all the islands that have been surveyed. For reasons explained, we call these *cloaking* metrics, for they tend to cloak or obscure the patterns. The chapter concludes by recommending that we examine each pair of species and their co-occurrences. Certainly, there are very large numbers of pairs in every one of the island systems we examine. We would expect there to be many pairs in the class of those that might appear to be usual in occurring as few times as they do. There are! There is a strong expectation, however, that ecologically related pairs will be especially numerous in that class. We test that idea in chapters 7 and 8.

How to Incorporate Constraints into Incidence Matrices

Definitions and Notation

We start by introducing some definitions and notations. We present the mathematical notation employed by previous researchers and used throughout our subsequent analyses. An explanation is given for two-dimensional matrices that are used to represent species on islands or sites, and for one-dimensional rows and columns within a matrix. We also define commonly used terms such as *null matrix, full null space, sample null space*, and what is meant by the term *representative*.

We use the standard matrix notation of mathematicians and physicists. Matrices are given in single bold letters. All the matrices we use have both rows and columns. Thus, $\mathbf{A}$ refers to a matrix and $a_{i,j}$ refers to the (i, j) element of $\mathbf{A}$ located in row i and column j. Hence, we can write $\mathbf{A} = \{a_{i,j}\}$ and this means that the matrix $\mathbf{A}$ is given by a collection of elements $a_{i,j}$. Usually the letter i refers to rows and j refers to columns. $\mathbf{A}$ is an n by m (or $n \times m$) matrix where n signifies the number of rows and m the number of columns. For our purposes, the bounds on i and j are $1 \le i \le n$, and $1 \le j \le m$, respectively.

The row sums and column sums of an $n \times m$ matrix $\mathbf{A}$ are denoted by the vectors $R = (r_1, r_2, \ldots, r_n)$ and $C = (c_1, c_2, \ldots, c_m)$, respectively. Henceforth, the row and column sums of the $n \times m$ matrix $\mathbf{A}$ are always be given as $\{(r_1, r_2, \ldots, r_n), (c_1, c_2, \ldots, c_m)\}$. If the row and column sums of $\mathbf{A}$ are identical, then a shorthand notation is used. For instance, if the row and column sums of a 2×2 matrix are both $(2, 1)$ then we write: $\{2(2, 1)\}$. If $\mathbf{A} = \{0\}$ then $\mathbf{A}$ is filled with all 0's. In the field of linear algebra, $\mathbf{A}$ is called the *null* matrix. Connor and Simberloff (1979) ignored this convention and used the term *null matrix* to refer to randomly created binary matrices. For historical reasons, we ignore the mathematician's notation here as well and adopt the term used by Connor and Simberloff (1979).

We also use the term *observed incidence matrix* to refer to the actual community observed in nature. The *full null space* is the complete collection of all unique null matrices satisfying certain constraints derived from the observed incidence matrix. Since creating the full null space is generally impossible, we most often create a *sample null space* that is a sample of the full null space. Presumably, the sample null space is representative of the full null space, but this turns out to be a very challenging and interesting presumption indeed—as we explain in the next chapter.

The Number of Null Matrices and the Effect of Constraints

If one abandoned the constraints altogether, there would be a universe full of random matrices with islands having no species and species occupying no islands. The simplest ecologically realistic way we could fill a matrix would be with exactly as many 1's as we observe in nature, or perhaps choose each entry with a given probability. In the previous chapter, we noticed that the Vanuatu matrix is about half full, so perhaps we would pick each entry with equal probabilities of having a particular species present or absent on that particular island. That model lacks a lot of ecological realism, but we nonetheless encounter it in chapter 10 about food web patterns.

Consider a half full *4x4* matrix. If no constraints are employed, there are a total of (16!/(8! 8!)) = 12,870 unique random matrices. Without constraints, two of the random matrices in the null space can contain empty rows (A) or columns (B) or both (C), for example:

$$A = \begin{matrix} 1 & 1 & 1 & 1 \\ 1 & 1 & 1 & 1 \\ 0 & 0 & 0 & 0 \\ 0 & 0 & 0 & 0 \end{matrix} \qquad B = \begin{matrix} 1 & 1 & 0 & 0 \\ 1 & 1 & 0 & 0 \\ 1 & 1 & 0 & 0 \\ 1 & 1 & 0 & 0 \end{matrix} \qquad C = \begin{matrix} 1 & 1 & 1 & 0 \\ 1 & 1 & 1 & 0 \\ 1 & 1 & 0 & 0 \\ 0 & 0 & 0 & 0 \end{matrix}$$

Next, consider a matrix with row and column totals {(3,2,2,1),(2,2,2,2)}. How many unique null matrices satisfy the row and column constraints? Here are 12 examples of null matrices:

```
1 1 1 0    1 1 1 0    1 1 1 0    1 1 1 0    1 1 1 0    1 1 1 0
1 1 0 0    1 0 1 0    1 0 0 1    1 0 0 1    1 0 0 1    0 1 1 0
0 0 1 1    0 1 0 1    0 1 1 0    0 0 1 1    0 1 0 1    1 0 0 1
0 0 0 1    0 0 0 1    0 0 0 1    0 1 0 0    0 0 1 0    0 0 0 1

1 1 1 0    1 1 1 0    1 1 1 0    1 1 1 0    1 1 1 0    1 1 1 0
0 1 0 1    0 1 0 1    0 1 0 1    0 0 1 1    0 0 1 1    0 0 1 1
1 0 1 0    1 0 0 1    0 0 1 1    1 1 0 0    0 1 0 1    1 0 0 1
0 0 0 1    0 0 1 0    1 0 0 0    0 0 0 1    1 0 0 0    0 1 0 0.
```

In these 12 unique random null matrices, the first row remained unchanged. There are four ways of placing three 1's in the first row: 1110, 1101, 1011, and 0111. For each of these four, there are 12 unique random matrices. This gives rise to a total of 48 unique random matrices. Without the row and column constraints, there are about 268 times more unique random matrices as with constraints. Obviously, variation in the null space has increased.

Consider a 5x5 matrix with just five 1's in it. There are 25!/(20! 5!) = 53,130 unique random matrices. Thus, without further constraints, there is *factorial growth* in the number of unique random matrices as the number of species and islands increases. Again, the constraints limit this number. To see this consider a 5x5 *identity matrix* **I** whose row and column sums are $\{2(1,1,1,1,1)\}$ that has a single 1 in each of five rows and columns:

$$
\begin{array}{ccccc}
1 & 0 & 0 & 0 & 0 \\
0 & 1 & 0 & 0 & 0 \\
0 & 0 & 1 & 0 & 0 \\
0 & 0 & 0 & 1 & 0 \\
0 & 0 & 0 & 0 & 1
\end{array}
$$

The null space consists of all matrices satisfying the row and column constraints. How many unique solutions are there? There are five ways to place the first 1 in the first row. Once the first 1 is placed, there are four ways to place a 1 in the second row, and so on. There are exactly 5! = 120 solutions.

What happens when we constrain the number of 1's, and either the row or column sum but not both? Clearly, the number of unique random matrices exceeds the number of matrices when both constraints are satisfied.

The Hypergeometric Distribution

Atmar and Patterson (1994), Dale (1984, 1986, 1988), Patterson and Atmar (1986), Pielou (1978, 1977, 1975), Pielou and Routledge (1976), and

Underwood (1978) and others chose to satisfy either the row constraint or the column constraint but not both to construct a random sample of communities. Consider this approach in more detail. Suppose the observed incidence matrix has row and column sums of $\{2(2, 1, 1)\}$. You will recall there are five unique matrices. Suppose we ignore the column sum constraint. In addition to

$$
A = \begin{matrix} 1 & 1 & 0 \\ 1 & 0 & 0 \\ 0 & 0 & 1 \end{matrix} \quad
B = \begin{matrix} 1 & 0 & 1 \\ 1 & 0 & 0 \\ 0 & 1 & 0 \end{matrix} \quad
C = \begin{matrix} 1 & 0 & 1 \\ 0 & 1 & 0 \\ 1 & 0 & 0 \end{matrix} \quad
D = \begin{matrix} 1 & 1 & 0 \\ 0 & 0 & 1 \\ 1 & 0 & 0 \end{matrix} \quad
E = \begin{matrix} 0 & 1 & 1 \\ 1 & 0 & 0 \\ 1 & 0 & 0 \end{matrix}
$$

more unique matrices are then possible

$$
F = \begin{matrix} 0 & 1 & 1 \\ 1 & 0 & 0 \\ 0 & 0 & 1 \end{matrix} \quad
G = \begin{matrix} 1 & 1 & 0 \\ 1 & 0 & 0 \\ 0 & 1 & 0 \end{matrix} \quad
H = \begin{matrix} 1 & 1 & 0 \\ 0 & 1 & 0 \\ 1 & 0 & 0 \end{matrix} \quad
I = \begin{matrix} 1 & 0 & 1 \\ 0 & 0 & 1 \\ 1 & 0 & 0 \end{matrix} \quad
J = \begin{matrix} 1 & 1 & 0 \\ 1 & 0 & 0 \\ 1 & 0 & 0 \end{matrix}
$$

and there are clearly others. The collection of unique random matrices is larger and the variation is greater. Note that in the absence of constraints, some islands can have no species while others have three.

Were both the row and column constraints abandoned and only the number of species present in the observed incidence matrix conserved, in this case four, this amounts to asking: how many ways can four species be placed in nine places?

This is, by now, the usual combinatorics question, and the simple answer is $9!/(5! \ 4!) = 126$. The null space has increased from having five unique members to having 126 unique members, and this example is only *3x3*. Variation in the null space has increased dramatically. Imagine the number of sample null matrices that might exist for Vanuatu were no constraints employed: $1,568!/(888! \ 680!)$. The number is truly astronomical.

Clearly, the row and column constraints act to structure the matrix and "constrain" possible random matrices, limiting the total possible number that exists. This limitation, however, is not a great concern. One can find truly amazing numbers of unique matrices for quite small samples—as we will soon see. (If one abandoned the constraints altogether, there would be a universe full of random matrices, with islands having no species and species occupying no islands.)

The Three Ecological Constraints Proposed by Connor and Simberloff in Their Studies of Birds and Bats on Islands

What Connor and Simberloff wrote in their abstract—this time in full— was: "We will show that every assembly rule is either a tautological con- sequence of the definitions employed, a trivial logical deduction from the stated circumstances, or a pattern which would largely be expected were species distributed randomly on islands subject to three constraints: (1) each island has a given number of species; (2) each species is found on a given number of islands; (3) each species is permitted to colonize islands constituting only a subset of island sizes." (We have reformatted this quo- tation for future use by labeling the constraints.)

We have already discussed how unnecessarily provocative the first part of the abstract is. Their important insight is what they called their *null model*. It is a means to test questions about pattern, and they provide the model constraints to do so.

The calculation involving two cuckoo doves in chapter 3 satisfied con- straint 2. Species *M. mackinlayi* and *M. nigrirostris* occur 15 times and 5 times, respectively, and 21 islands have neither. Unfortunately, constraint 1 says that each island has a given number of species, and it is this rule that Connor and Simberloff ignored in their simple calculation, by force of circumstance. By only considering the doves themselves and ignoring all other species, Connor and Simberloff violate constraint 2. This is no trivial concern. For when one employs both constraints 1 and 2 in the complete data set, as we do in chapter 6, the chance of finding the two species together falls from 1 in 39 (insignificant), to roughly 1 in 10,000 (highly significant).

Inspection of the map of the Bismarck Archipelago shows that the islands differ greatly in size, and the two tables show the corresponding ranges of species numbers—they range from 4 to 127. Any null hypothesis that did not constrain the numbers of species on all islands allows the co- occurrence of 5 fruit doves and 4 cuckoo doves, or 9 species in total— and 5 species more than some of the smaller islands contain of all species. Equally, if islands are populated under a random allocation, some will have too many species, some too few, and perhaps very few with exactly the number observed in nature.

Similar arguments apply to the species totals—some species are good at dispersing, others are not. By extension, one might argue that one should restrict the incidences, because some, but not all, islands may be out of reach for some species. These arguments have a hidden problem. Some species only occur on islands with few species—the supertramps. Yes, they might reach such islands because they can disperse well and the islands may be remote, but why don't they occur on islands with more species? Perhaps their congeners exclude them. As we shall explain presently, constraints 1 and 3 are contentious. Simply, to analyze patterns of coexistence we need more complete methods than those used in the straightforward cuckoo dove model presented in chapter 3.

We now examine these constraints in more detail. As pointed out above, some authors later chose to ignore some or all of these constraints—much to their own peril. Indeed, even Connor and Simberloff (1979) found it difficult to create random communities satisfying these three constraints. To their credit, they refused to abandon the constraints and used them to publish the now-classic paper that set off the debate.

Incidence

Tables 2.1 and 2.2 show the islands and how many species occur on each island. Throughout, we have followed the convention that the term *small island* refers to islands with few species and *large island* refers to islands with many species. Generally this is true because physically large islands likely have more habitats than small islands and so support more species. Hence, when authors discuss "large islands" they are referring to islands with comparatively more species than are found on "small islands" that have fewer species.

Each species incidence relates the total number of species found on each island to the absence or presence of the particular species in question.

Figure 2.9 has already shown the distribution of the *Myzomela pammelaena* and *M. sclateri*—both "supertramps" found only on "small" islands (small in the sense that the islands have small areas but also few species—no more than 58 and 54, respectively). Another species, *Myzomela cruentata*, occurs on the largest islands, with Dyaul having 50 species and the other four islands 60 or more.

Incidence constraint 3 requires that species in the sample null communities are constrained to occur on the same size island as in the observed community. For instance, an island or site endemic that occurs on the most species-rich islands must always do so in each sample null community.

Thus, the incidence constraint limits the total number of random communities satisfying all three constraints.

With the identity matrix, the incidence constraint has no effect. If one adds another presence to the matrix, however, then the incidence constraint comes into play. To see this, consider the following example.

	Islands				
Species 1	1	0	0	0	0
Species 2	1	1	0	0	0
Species 3	0	0	1	0	1
Species 4	0	0	0	1	0
Species 5	1	0	0	0	0

The incidence constraint says that species 1, 2, and 5 must occur on the most populated island, island 1, in all null matrices that make up the full null space. Therefore, the first column is fixed. No other species can be placed in the first column except 1, 2, and 5. However, the remaining 1's can be placed arbitrarily according to the row and column constraints. Again, the incidence constraint acts strongly to limit the number of possibilities.

What about larger examples? Trivially, if a matrix of any size is completely filled with 0's or 1's, then only a single null matrix exists. This is because there is no flexibility to have any other arrangements of 1's and 0's—with or without constraints. If an observed occurrence matrix is of some ordered form, then the number of matrices is also limited. Consider a 5x5 upper triangular matrix with 15 1's

$$
\begin{array}{ccccc}
1 & 1 & 1 & 1 & 1 \\
0 & 1 & 1 & 1 & 1 \\
0 & 0 & 1 & 1 & 1 \\
0 & 0 & 0 & 1 & 1 \\
0 & 0 & 0 & 0 & 1 \\
\end{array}
$$

If the row and column constraints must be satisfied, how many unique random matrices exist? Clearly, there is only one matrix that satisfies both constraints. In each null matrix, the first row must be fully occupied. There is only one way to achieve this pattern. After the first row is filled, the first column constraint is also satisfied. Hence, there is only one way to satisfy the second row constraint. The same is true with the remaining rows of the matrix when both constraints are applied. The null space consists of a single matrix and that matrix is **U**.

Now consider a 5x5 matrix with 15 1's randomly placed as follows:

```
1 0 1 0 1
0 1 0 1 1
1 0 1 0 1
0 1 1 1 0
1 0 1 0 1
```

The row and column sums are $\{5(3, 3, 3, 3, 3)\}$. The row and column constraints require that each row and column sum be 3. This can happen in many ways.

Generally, the row and column constraints derived from the observed incidence matrix limit the total number of null matrices. As the observed incidence matrix approaches being empty or full, the total number of random matrices decreases. If the observed incidence matrix is roughly half full and there appears to be no particular order, then the number of possible random matrices approaches a maximum when the matrix is half full. A graph of the total number of unique null matrices that satisfy both row and column constraints as a function of the number of 1's in the observed incidence matrix rises and falls symmetrically (fig. 4.1).

Now consider a *7x8* example with row and column sums $\{(3, 2, 2, 2, 2, 1, 1),$ $(2, 2, 2, 2, 2, 1, 1, 1)\}$. One example null matrix is

```
1 1 1 0 0 0 0 0
0 1 0 1 0 0 0 0
0 0 0 0 1 0 0 1
1 0 1 0 0 0 0 0
0 0 0 0 0 1 1 0
0 0 0 0 1 0 0 0
0 0 0 1 0 0 0 0
```

Using a variety of generation methods discussed in the next chapter, we generated 1,271,190 unique null matrices that satisfied the row and column constraints.

By adding one row and one column, we created an *8x9* matrix with row and column sums $\{(4, 3, 3, 2, 2, 2, 1, 1), (4, 3, 2, 2, 2, 2, 1, 1, 1)\}$. Again, by using a variety of generation methods, we generated 93,833,092 unique null matrices that satisfied both constraints. The addition of a partially filled single row and column increased the number of unique null matrices by a factor of nearly 100.

In these examples, the incidence constraint, if applied, would act to limit the number of possible null matrices. The incidence constraint never acts to increase the number of possible random matrices. The incidence constraint acts more strongly when there are greater differences in the column sums because the constraint forces species to occur on islands with similar species numbers. The row and column constraints also limit

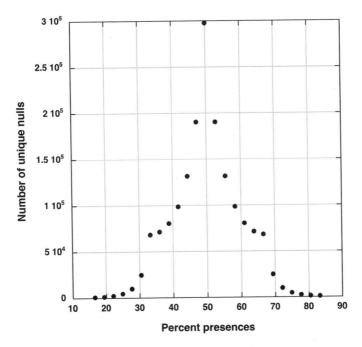

FIG. 4.1. The number of unique random matrices in a *6x6* randomly populated observed community as a function of the number of presences. Each null matrix satisfies the row and column constraints.

the number of possible unique null matrices, but less strongly. It should be clear that other constraints are possible, depending on the situation. For instance, for an analysis of species along an ecological gradient (chap. 8), the range spanned by each species might be contiguous. Such a constraint strongly limits the number of possible unique null matrices. Observed co-occurrence patterns, however, are invariant when rows and columns are shuffled. At a minimum, an analysis of patterns formed by species co-occurrences should use both the row and column constraints.

Most archipelagos consist of more than 8 species and have more than 9 islands. The Galápagos Islands have at least 17 islands for which we have survey data; the Bismarck Archipelago has 41 major islands with more than 150 species of birds; the Solomon Islands are a collection of more than 140 islands and 140 species of birds. Vanuatu has 56 resident species of land birds distributed across 28 islands. For these examples, the observed incidence matrices are approximately half full. As we have seen, one can only imagine the total number of unique null matrices that satisfy the row and column constraints. Ralph Selfridge, formerly a computer scientist at the University of Florida, estimated that the total number of unique random matrices of Vanuatu exceeds 10^{40}. If a computer could create one null matrix in a millionth of a second, then in a billion years only 3×10^{10} would be created—a very tiny fraction of the 10^{40} unique null matrices. It is not possible to find all of them, and therefore *sampling* the null space is essential.

Sampling the null space is the only way to analyze complex ecological communities. The *sample null space* refers to the collection of unique random matrices created to be *representative* of the *full null space*, the complete collection of all unique random matrices that satisfy the row and column constraints. By "representative"—described in more detail below—we mean that the sample null space is a collection of unique uniform-randomly created null communities from the full null space. Ideally, if the full null space was known a priori, then it is a simple matter to choose a sample without replacement uniform-randomly. However, if the full null space were known, there would be no need to do so.

Connor and Simberloff (1979) used only ten sample null matrices. Each null matrix satisfied the row and column constraints. Their fears that very few null matrices existed have proved utterly unfounded. In fact, the problem is far more severe and problematic than they ever imagined. Indeed, we show that creating a sample null space that is truly "representative" of the full null space is a formidable challenge.

Ideally for any problem, the full null space could be computed according to agreed-upon constraints, and the co-occurrence pattern of every possible species pair, trio, guild, genus, family, or collection could be tested. There has been much progress in this pursuit. For "small" problems, the goal of generating the full null space has been achieved, and "small" continues to grow with each advance in computing power. For those problems whose full null space cannot be generated, one must compute a sample null space that is somehow "representative" of the full null space.

Why Constraints? And What Does "Representative" Mean?

Why should the null space be constrained at all? The answer is that each null matrix must somehow be representative of the observed incidence matrix. (We discuss this in more detail in chap. 5.) When a metric is used to compare a pattern in the observed incidence matrix against the "average" pattern derived from the sample null space, we must have some assurance that we are comparing the observed pattern to one that could have arisen by chance. In the absence of constraints, the possibility that the observed incidence pattern will prove to be significantly different from chance expectations increases simply because the observed pattern is compared to a greatly enlarged sample null space with extraordinary variation.

Imagine creating a random but representative community of an observed community that has five islands and just five endemic birds, one endemic bird for each island, and each island having just one bird. A simple arrangement gives the familiar 5x5 identity matrix. There are many choices of random matrices to create a sample null space, but the question is which random choices are representative of the observed community and which are not. Authorities now are in agreement that a random community consisting of, for example, all five birds on one island, and hence four islands with no birds is not representative of the observed community.

A simple way to describe the observed community is to write: there are five birds, each on a single island, and each island has just one bird. This statement of the observed community precisely specifies the constraints needed to create a representative null space, namely, the observed row and column sums of the observed community must be satisfied in each random community. Each island in each random community has one-and-only-one bird of the five birds, and each bird is on one-and-only-one island. With such constraints, the number of random communities is less than the number of unconstrained communities. The importance of this fact should not be underestimated for the following reason. The value of a metric, any metric, computed from the observed community is compared to the distribution of values computed from each member of the sample null space. If the sample null space has fewer constraints—for instance, only the row sums are conserved but the column sums are not—or is unconstrained, then the variability of the metric values increases. Therefore, the observed value of the metric is more likely to exceed chance expectations. The opposite is also true.

If more constraints are imposed, the variability in the null space decreases and hence the more likely it is that the observed value of the metric will not exceed chance expectations. The limit would be a list of constraints that eliminates all variability, leaving but one member of the null space, namely, the observed community.

Thus, in the absence of constraints, many more observed patterns will be found to be unusual because variation in the null space is so great. As additional and more stringent constraints are applied, variation in the null space decreases so that more patterns in the observed incidence matrix become less unusual. When yet more stringent constraints are applied, the null space becomes overly constrained, variation disappears, and the null matrices all closely resemble the observed incidence matrix. In this case, no patterns will be unusual.

Failure to understand the effects and consequences of the constraints led Fox to write: "The major debate on the guild assembly rules comes down to whether one should maintain the distribution of species richness across sites . . . or whether one should retain the distribution of each species' frequency of occurrence across sites. . . . The sensible course would seem to be to agree to differ on these points and get on with research activities that advance our knowledge more" (1999, 40).

Singling out Fox is unfair as others before him also did not fully appreciate the consequences of null matrices unfettered by constraints. Other authors used null matrices to analyze communities (Gotelli and Graves 1996; Gotelli and Abele 1982; Bolger et al. 1991; and many others), and some such as Gotelli and Graves (1996) also recommended that row and/or column sums act as probabilistic constraints and should be allowed to vary. Diamond and Gilpin (1982) even used fractions to create null communities. Obviously, we disagree with these views.

Summary

This chapter describes the first step of a two-step process used to analyze ecological communities. The first step is to create a set of random communities that satisfy the row and column constraints needed to ensure that the random communities resemble the observed incidence matrix in ecologically important ways. Here, we show that the numbers of possible random matrices quickly become vast for incidence matrices of the size of those we wish to analyze, but that sensible ecological constraints limit those numbers. Despite the constraints, one can only create a sample collection of random communities and thus begging the question of how representative they are of the complete set. The second step of the process, described in the chapter 6, is to create a metric to measure something ecologically interesting about the observed incidence matrix and then to see how the metric values are distributed statistically across the sample of random communities. The next step, however, is to create random communities.

How to Fill the Sample Null Space

Nothing makes sense in the analysis of species co-occurrences except in light of binary matrices. — Anonymous

There are a limited number of such matrices. — Connor and Simberloff 1979

There are *a quite limited number of possible matrices*. — Wilson 1987

For reasons presented previously, creating the full null space at this time is computationally impossible for the situations we find interesting. A sample null space is all that we can hope to create. The problem, as noted previously, is that the sample null space must be representative of the full null space. What does *representative* mean? The sample null space is representative of the full null space when properties of the sample null space are identical to properties of the full null space. That is, for whatever metric one chooses, the sample null space must show the same distribution as the full null space.

For example, suppose that the full null space contains 360 unique null matrices. Suppose further that the metric values derived from each null matrix range from 1 to 11. We can then plot the metric values versus the frequency of each value and compare the plot to the same values derived from smaller sample null matrices each consisting of 90 samples (fig. 5.1). In this hypothetical example, sample null space 1 is similar to the lower end of the metric values but fails to sample the higher metric values and so is not representative of the full null space.

Sample null space 2 more closely resembles the full null space. The problem is, however, that for real problems we do not know the full null space. What assurances do we have that the sample null space we create is representative of the full null space?

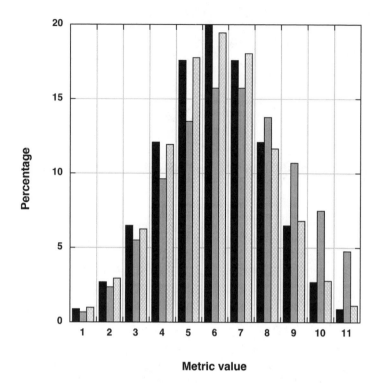

Metric value

FIG. 5.1. A comparison of the full null space and sample null spaces. The full null space consists of 360 unique null matrices. Two sample null spaces each containing 180 unique null matrices are also given. Sample null space 2 is representative of the full null space because the frequency of metric values is identical to those in the full null space. This example is hypothetical and not based on an actual example. Black bars indicate the number of times the metric value was found in the observed community; dark gray and light gray bars indicate the number of times each metric value is found in each of two sample null spaces.

Null Space Creation Algorithms

An algorithm is a repeatable procedure used to produce something. A typical cookbook contains many algorithms. An instruction manual such as that used to assemble a model airplane usually gives a list of steps, an algorithm, that when followed results in proper assembly of the model. There are two classes of algorithms used to create the sample null space: (1) swap algorithms, and (2) construction algorithms.

These approaches are fundamentally different although in theory both can produce "representative" sample null spaces. "In theory" is very different from "in practice," however, and an algorithm can begin as a construction algorithm and then employ a swap algorithm to create a sample null space member. Another remaining obstacle to the proper analysis of species co-occurrences is the creation of a sample null space that can be proven to be representative of the full null space. If every member of the null space is needed, enumeration algorithms are most efficient because each member of the null space is produced systematically from the first member to the last.

Enumeration Algorithms

Enumeration algorithms generate every member of the null space one after the other, systematically. Enumeration algorithms are thus *exhaustive* algorithms. Typically, enumeration algorithms generate sequential members that are nearest neighbors and hence most similar. Thus, enumeration algorithms cannot be terminated until the full null space is generated; otherwise, part of the full null space would not be sampled. For small problems, enumeration algorithms are useful. For typical problems of interest to ecologists that have huge full null spaces, we cannot use enumeration algorithms.

An enumeration algorithm can use either swap algorithms or construction algorithms to populate the full null space. A construction enumeration algorithm might start from a matrix of o's and fill it some way while satisfying the constraints. The algorithm then retreats one step from the solution and proceeds from this advanced state to create all other neighboring solutions. After that, the enumeration algorithm retreats one step back and again proceeds forward to find other null matrices. Each step finds representatives of the null space that are most similar because a retreat is one step backward, leaving much of the initial advanced state identical. Premature termination, for instance after a certain number of null communities is achieved, would leave a portion of the full null space without representation in the sample null space.

Enumeration algorithms suffer from two fatal flaws: they are combinatoric in nature, and they are systematic. That is, one creates every possible null matrix systematically and then tests it against the constraints. If the constraints are satisfied, then the null matrix becomes part of the sample null space. In theory, such a procedure produces every null matrix with equal probability. In practice, enumeration algorithms only work for problems with small full null spaces.

Recall in chapter 4, the *4x4* example with row and column sums {(3, 2, 2, 1), (2, 2, 2, 2)}. There are 8 presences and 16 positions. There are thus 16!/(8! 8!) = 12,870 ways of placing 8 1's in 16 positions. In this example, just 48 solutions satisfy the row and column constraints. Generally, as the size of the observed incidence matrix increases, the number of possible matrices increases very rapidly while the number of actual matrices grows more slowly. The result is that a lot of work is needed to find the appropriate random matrices. In this example, there is roughly one appropriate random matrix for every 268 possible matrices.

Now consider a *5x5* example with row and column sums {5(3, 3, 2, 2, 2)}. There are 12 1's and 25 positions. There are 25!/(13! ×12!) = 5,200,300 ways of placing 12 1's in 25 positions but there are only 264 matrices in the sample null space. There is now just one matrix for every 19,698 possible enumerations. The full null space doubled in size from the previous example, but the number of possible matrices to check increased by more than 400 times. One more example will convince even the most skeptical among us that enumeration algorithms are useless in practice.

Consider a *6x6* example with row and column sums {6(3, 3, 3, 3, 3, 3)}. The number of 1's is 18, and there are 36 positions. There are thus 36!/(18! × 18!) = 9,075,135,300 ways of placing 18 1's in 36 positions. There are just 7,650 solutions. There is now just one solution for every 1,186,292 possible enumerations. The full null space increased thirtyfold in size from the previous example, but the number of possible solutions that one must check increased by about 175 times.

Now imagine the Vanuatu example with 56 species on 28 islands with about half the number of possible positions occupied. Clearly, we cannot use an enumeration algorithm to populate the sample null space. How then is a sample null space created that is much smaller than the full null space but still representative of the full null space? The idea is to avoid the systematic nature of enumeration algorithm and introduce a randomization procedure. We can employ both swap and construction algorithms and combinations of the two.

Swap Algorithms

A construction algorithm that creates null matrices starting with a matrix of zeros proved elusive for Connor and Simberloff (1979) as they described a *matrix-filling process* that "cannot be completed." They resorted to an alternative algorithm to construct Vanuatu null matrices to avoid a "hang up." Since they were unable to provide a construction method, Connor and Simberloff instead began with the observed incidence matrix. They started by placing the actual Vanuatu incidence matrix in what they called "a canonical form" by putting rows and columns in "echelon form" as described in finer text in their appendix. Echelon and canonical forms were simply reorderings of rows and columns and fundamentally changed nothing. Think of the rows ordered from the most populated to the least populated and the same for the columns. This is *canonical ordering*.

In the Vanuatu observed incidence matrix, their algorithm searched for patterns of the submatrix form:

	Islands			Islands	
	l	m		l	m
species i	1	0	or	0	1
species j	0	1		1	0

where species i and j and islands l and m were not necessarily in adjacent rows and columns, respectively. Whichever pattern was found, it was transposed into the other. In this way, row and column sums remain unchanged.

Returning to the *3x3* example with row and column sums {2(2, 1, 1)}, recall there were five matrices, **A–E**, that satisfied the constraints:

$$A = \begin{matrix} 1 & 1 & 0 \\ 1 & 0 & 0 \\ 0 & 0 & 1 \end{matrix} \quad B = \begin{matrix} 1 & 0 & 1 \\ 1 & 0 & 0 \\ 0 & 1 & 0 \end{matrix} \quad C = \begin{matrix} 1 & 0 & 1 \\ 0 & 1 & 0 \\ 1 & 0 & 0 \end{matrix} \quad D = \begin{matrix} 1 & 1 & 0 \\ 0 & 0 & 1 \\ 1 & 0 & 0 \end{matrix} \quad E = \begin{matrix} 0 & 1 & 1 \\ 1 & 0 & 0 \\ 1 & 0 & 0 \end{matrix}$$

Note that **A** can be transformed into **B, D,** or **E** by performing a single swap. Two swaps are required to turn **A** into **C**. Note also that further swaps perform cyclic operations on **A–E**. Using this example, suppose that we start with **A** each time and use one swap to generate the sample null space. Then the null space will consist only of **B, D,** and **E**. Frequently, a program uses a priori a fixed number of swaps to generate a sample null. If the same starting matrix is used for each null creation, the result is the same as just observed. In this *3x3* example the effect of applying more swaps is to regress to previously generated null matrices. Moreover, the sample null space will have small variation because all the members will be more or less the same "distance" away from the starting matrix, in this case the observed incidence matrix. As we show later, **A, B, C, D,** and **E** are not equally likely to be produced.

Swaps do not alter the row or column sums but do alter the incidence constraint (chap. 2) if species i and j populate different-sized islands. Connor and Simberloff (1979) applied all three constraints, so additional care had to be exercised. For Vanuatu, results were presented using all three constraints though there was little difference with or without the incidence constraint.

According to Connor and Simberloff, after several swaps (or transpositions) to the observed incidence matrix, the null matrix was included in the sample null space. It is not clear just how many transpositions were applied. Fewer swaps would give rise to null matrices being relatively similar. Thus, it is hardly surprising that the metric value derived from the sample null space would differ much from that derived from the observed incidence matrix. Indeed, this is what stuck out most when we saw the first figure in Connor and Simberloff (1979) and given here in chapter 3. The fit was just too good to be true; something had to be wrong, didn't it?

The Case for Swaps

Brualdi (1980) showed that, with a certain number of well-chosen swaps, any representative matrix in the null space could be transformed into any other representative matrix in the null space. In effect, one could generate, in principle, every member of the full null space by applying a small number of well-chosen swaps. For problems of interest to ecologists, this procedure is not possible because of the enormous size of the full null space. We can easily calculate the number of swaps necessary to go from one sample null matrix to another. The *distance* from one sample null matrix to another is simply the number of positions that differ between the two null matrices divided by four.

For instance, in the *3x3* example with five unique solutions:

$$
A = \begin{matrix} 1 & 1 & 0 \\ 1 & 0 & 0 \\ 0 & 0 & 1 \end{matrix} \quad
B = \begin{matrix} 1 & 0 & 1 \\ 1 & 0 & 0 \\ 0 & 1 & 0 \end{matrix} \quad
C = \begin{matrix} 1 & 0 & 1 \\ 0 & 1 & 0 \\ 1 & 0 & 0 \end{matrix} \quad
D = \begin{matrix} 1 & 1 & 0 \\ 0 & 0 & 1 \\ 1 & 0 & 0 \end{matrix} \quad
E = \begin{matrix} 0 & 1 & 1 \\ 1 & 0 & 0 \\ 1 & 0 & 0 \end{matrix}
$$

The number of positions that differ between **A** and **B** is four. Thus, only one swap is needed to generate **B** from **A**. **C** differs from **A** in six positions, thus two swaps are required to go from **A** to **C**. Note that by applying one additional swap to **C**, any other null matrix that is a distance one away from **C** will be generated: **D** or **E**. Thus, if one applies three swaps to **A**, **C** will never be generated. Note that **B**, **D**, and **E** can all be reached from **C** by just one swap. This fact has serious implications for swapping algorithms, as we will see below.

For the Vanuatu data set, Gotelli (2000) applied 10,000 initial swaps to the observed incidence matrix and then retained a sequential set of 1,000 swaps to form the null set (Gotelli and Entsminger 2003). The first 10,000 initial swaps *displace* the observed incidence matrix. Their sample null space was then created by applying *a single swap* to the previous sample null space member. Each successive null space member is as minimally different from the previous member as is possible. Obviously all the null space members will appear alike, and there will be minimal deviations in the results. In the algorithm they used (EcoSim; Gotelli and Entsminger 1999), each subsequent null space member is entirely dependent on the previous member so that the sample null space is in no way representative of the full null space.

Roberts and Stone (1990) also used a swapping method. After 100,000 swaps, Roberts and Stone then subjected the resulting matrix to J' random interchanges, then J'' random interchanges, and so forth. They chose ad hoc values of J, which was an excellent idea as it avoided using a fixed number of swaps. However, Roberts and Stone (1990) provided neither the number nor the bounds on J. Once again, the observed incidence matrix was displaced and then perturbed, but this time by an unknown random number of swaps. Presumably, these additional random swaps act to increase the distance between neighbors around the displaced matrix. Again, the question is whether the sample null space is in any way representative of the full null space. Roberts and Stone (1990) provided no such evidence. But we don't want to pick on them as they were not alone, methodologically speaking. In fact, we return to the work of Roberts and Stone in the next chapter because they also offered a metric by which to test for patterns in the observed community.

Referring to the null matrices of Connor and Simberloff (1979), Diamond and his colleagues were quick to point out, quite correctly, that if the *2x2* swaps were performed only a few times, one could not expect to see large changes in the numbers of co-occurring pairs of birds. Indeed, each step perturbed the numbers of occurring/nonoccurring pairs only slightly from the observed incidence matrix and, as noted by Diamond and Gilpin (1982), was not random anyway. In other words, the null matrices produced by Connor and Simberloff (1979) were just slight perturbations of the observed incidence matrix. Consequently, any results they produced would be spurious—as we show in the next chapter.

Although Diamond and Gilpin's (1982) arguments were predominantly ecological, they offered alternate methods for null matrix construction. One method proposed by Diamond and Gilpin (1982) used fractional values in the sample null matrices. For instance, instead of placing 0's or 1's to indicate absence or presence, respectively, the authors suggested using fractions to indicate probabilities of occurrences.

The reason that alternative methods were proffered for null matrix creation was the difficulty researchers found in creating null matrices whose row and column sums matched those of the observed incidence matrix. Nevertheless, much to the credit of the opposing camps such as Connor and Simberloff (1983) and Diamond and Gilpin (1982, 1983), both agreed on the need to satisfy the row and column constraints and the desirability of an unbiased method to generate null matrices. The problem was that no unbiased method had been identified.

Construction Algorithms

Wilson realized the value of a construction method to create null matrices, though he too believed there to be "a quite limited number of possible matrices." He suggested allocating "at each stage either to the island remaining that has the highest number of unallocated occurrences, or to the species remaining with the highest number of unallocated occurrences, whichever is the highest number" (1987, 580). As Wilson claimed, this produces null matrices but was systematic because each island was populated in nearly the same sequence (though most likely with different species) for each null matrix.

There is a simple way to obtain a purely uniform random sample null space. Unfortunately, the algorithm does not scale to larger problems because, as we have seen, the number of null matrices satisfying the constraints is dwarfed by the number of matrices that have the same amount of fill (the number of 1's, presences) but do not satisfy the constraints. Fundamentally, this is why other algorithms have been pursued.

The Knight's Tour

In 1998, Mike Moulton, Ralph Selfridge, and Sanderson (Sanderson et al. 1998) created a fast and efficient construction method called the "knight's tour," which started with a matrix entirely of 0's and populated it while satisfying the row and column constraints. The objective of the knight's tour is to determine the sequence of moves that allows a knight to move across a chessboard, touching every square just once. To begin, one places a knight randomly on the chessboard. The knight advances until it either touches every square just once or cannot move because it would touch a square twice. In the latter case, the knight is backtracked once to the previous square (commonly referred to as a *decision point*) and then advanced forward until it cannot move or it reaches completion. The knight continues to move forward until it cannot move and then backtracks to the most forward previous square until all alternatives are exhausted, in which case the knight is backtracked again. In this way, the knight traverses every square just once. Additional traverses can be generated by returning to a previous decision point—as Zaman and Simberloff (2002) suggested—or by starting over from a matrix of all zeros. If the former, then subsequent matrices would not be independent. The latter guarantees independent results.

The method was *recursive* because it successively added 1's to the matrix of 0's. Occasionally the algorithm had to reject the placement of a 1 as a dead end, meaning that the placement violated either a row or column sum constraint. In such cases, the procedure was to backtrack to a previous step and then proceed forward by placing a 1 in a different position. Such a method would not have been possible without advances in the computing power of desktop computers.

Simply, Sanderson et al. (1998) developed a fast and efficient recursive algorithm that created independent random matrices from a matrix of zeros that exactly satisfied each row and column constraint. They based their algorithm on well-understood computing science techniques. It therefore avoided the ad hoc procedures (e.g., Roberts and Stone 1990, which Manly 1995 noted, and those employed by Wilson 1987) associated with starting with the observed incidence matrix and then deciding when to stop the perturbation process. The knight's tour algorithm terminated when all row and column constraints were satisfied and then restarted from a zero matrix to generate additional random matrices. This process assures independence by starting from a zero matrix. Thus, the use of a fast construction algorithm enables populating the sample null space with independent null matrices. There is no proof, however, that each representative contained in the null space has an equal probability of being generated. In fact, each random matrix did not have an equal probability of being created, but more on that later.

A related problem also based on the game of chess is the "eight queens" problem. Here the goal is to place eight queens on a chessboard so that no two are attacking each other and such that all squares are attacked at least once. Both the knight's tour and the eight queens algorithms are solved problems in the field of computing science. To see how one applies the backtracking algorithm to the creation of null matrices, consider as an example two species occupying five islands with row constraints and column constraints $\{(3,4),(2,1,1,2,1)\}$.

Starting from a matrix of zeros, a species presence is established by uniform randomly choosing a row index i and a column index j and filling it with a 1, the presence of species i of island j. Here index $(1, 1)$ was randomly selected first, then index $(2, 3)$ as indicated by the superscript numbers:

3	1^1	0	0	0	0
4	0	0	1^2	0	0
	2	1	1	2	1

Step 2 matrix; bold numbers to the left and bottom are the target row and column totals.

Proceeding with the filling process, suppose index $(2, 1)$ is chosen next and filled, followed by index $(1, 4)$, and then $(2, 5)$, resulting in the following:

3	1^1	0	0	1^4	0
4	1^3	0	1^2	0	1^5
	2	1	1	2	1

Step 5 matrix

Note that this matrix does not (yet!) violate any constraint.

During null matrix construction, unfavorable states may occur that require backtracking to a previous favorable state. For instance, suppose at the next (sixth) step, we chose index $(2, 2)$ to fill. Clearly, this is possible and it violates no constraints.

3	1^1	0	0	1^4	0
4	1^3	1^6	1^2	0	1^5
	2	1	1	2	1

An unfavorable step 6

This placement creates an unfavorable state because there is no way to fill another state and satisfy row constraint 1 and column constraint 4 simultaneously. The knight's tour algorithm has reached an impasse.

The solution requires backtracking to the previous favorable state and then advancing. We restore index $(2, 2)$ to zero and do not consider it further. Backtracking occurs to step 5 (above), and we chose another index randomly again so as not to violate any row or column sums. Here, index values $(1, 2)$ and $(1, 4)$ are viable choices. Suppose index $(1, 4)$ is chosen.

3	1^1	0	0	1^4	0
4	1^3	0	1^2	1^6	1^5
	2	1	1	2	1

Step 6

By uniform randomly selecting another (i, j) location for step 7—the choice of index $(1, 2)$—we can eventually complete the matrix.

3	1^1	1^7	0	1^4	0
4	1^3	0	1^2	1^6	1^5
	2	1	1	2	1

Step 7, the complete matrix

In this way, the algorithm proceeds through a progression of favorable and unfavorable states, successively advancing and backtracking until it achieves a solution. Once a solution is reached, we compute the metric, and the knight's tour algorithm begins again from a matrix of all 0's.

After the computer program generated a sample null matrix, the program computes the metric such as the number of co-occurrences of each possible species pair described below. Sanderson et al. (1998) used the random number generator supplied with the APL compiler. The so-called *congruential generator* was typical of most random number generators since it produced a "flat" distribution between 0 and 1 with a cycle period in excess of 10^{10}. This program chose a random number, scaled it to between 1/2 and the number of rows + 1/2, and computed the nearest integer that then became the row index. The procedure then repeats to obtain a column index. Each (i, j) position in the matrix had an equal probability of being chosen. As we were to discover, however, choosing each index uniform randomly during null construction does not guarantee that a sample null space is a uniform random sample of the full null space. Moreover, this was also true of swap algorithms.

The row and column constraints determine, in a complex way, how to construct the null matrix even though the matrix is filled by uniform random choices. For instance, in a simple *2x2* case let the row and column sums be both $(1, 1)$. Once the first 1 is placed (and this is uniform random), all subsequent positions are determined by the row and column sums. Obviously, a 1 must be placed on the opposite diagonal, but this eventually occurs randomly. So, although all (i, j) entries are chosen uniform randomly, the row and column sums exert increasing pressure as the algorithm proceeds. The 1's initially chosen are rarely changed, whereas the ending 1's are juggled until the row and column sums are satisfied.

Creating a Uniform Random Sample Null Space

Creating a sample null space with the same row and column sums as those found in the observed community is straightforward. We must address a more difficult question: Is the sample null space representative of the full null space?

Suppose the sample null space unevenly samples the full null space because of inherent biases in the sampling algorithm. This problem is simple to investigate but a very challenging and formidable one to solve. A simple example demonstrates that *unbiased* swapping and construction methods do not sample the full null space uniform randomly. There is only one surefire way to create a sample null space that is a *uniform random sample* of the full null space. Unfortunately, this method only works when each member of the full null space is known, and if that's the case, why bother to create a sample null space?

To create a representative sample of M members of the full null space is simple: (1) generate each and every unique member of the full null space (perhaps by an enumeration algorithm); (2) label each member from 1 to N; and (3) choose M integers uniform randomly from 1 to N without replacement.

This procedure guarantees that the sample of M null matrices is indeed representative of the full null space. The problem is obvious: to derive the sample null space, every member of the null space must be known a priori. Generally and practically this is impossible.

A Simple 3x3 as an Example

Returning to the *3x3* example with row and column sums {3 (2, 1, 1)}, the full null space consists of five unique members:

$$A = \begin{matrix} 1 & 1 & 0 \\ 1 & 0 & 0 \\ 0 & 0 & 1 \end{matrix} \quad B = \begin{matrix} 1 & 0 & 1 \\ 1 & 0 & 0 \\ 0 & 1 & 0 \end{matrix} \quad C = \begin{matrix} 1 & 0 & 1 \\ 0 & 1 & 0 \\ 1 & 0 & 0 \end{matrix} \quad D = \begin{matrix} 1 & 1 & 0 \\ 0 & 0 & 1 \\ 1 & 0 & 0 \end{matrix} \quad E = \begin{matrix} 0 & 1 & 1 \\ 1 & 0 & 0 \\ 1 & 0 & 0 \end{matrix}$$

Construction Algorithm

This example begins with a *3x3* matrix of zeros (0's). A uniform random construction method starts by selecting an index (i, j) uniform randomly and then placing a presence in the index. Initially, each index has a probability of $(1/9)$ of being selected. The problem is that certain initial selections invariably lead to a single unique solution. For instance, an initial selection of the index $(3, 3)$ leads inevitably to the creation of solution **A**, just as a selection of $(3, 2)$ leads to **B**, $(2, 2)$ leads to **C**, and $(2, 3)$ leads to **D**. These same solutions, however, can also be created from other initial selections as well. For instance, an initial selection of $(1, 1)$ can lead to the creation of members **A**, **B**, **C**, or **D**. Note that no initial selection leads solely to **E**. Moreover, once the initial selection is made, the constraints begin to exert their control over the outcomes.

It is possible to enumerate and calculate the probability of occurrence of every solution **A–E**. Begin with a *3x3* matrix of all zeros. Note that one can fill any index chosen in the first step. After the first index is chosen, eight indices remain. These are subject to constraints, however. If a constraint is violated, the choice must be abandoned. In effect, this reduces the number of choices that can lead to a solution. For instance, suppose $(3, 3)$ was the initial choice and was filled. This choice prevents any further choices in the last column. Any other choices in the first two columns are viable. Thus, any of six index choices (three each in the first two columns) can be made with probability 1/6.

Following all the choices to a successful null construction subject to the constraints is possible in this simple example. By using this purely uniform random construction method, the probability of **A** occurring is $(1/9) \times (89/48)$; **B**, $(1/9) \times (92/48)$; **C**, $(1/9) \times (95/48)$; **D**, $(1/9) \times (92/48)$; and **E**, $(1/9) \times (64/48)$. These probabilities are not uniform; test runs of the computer program confirm these results.

In this example, there are only five unique null matrices in the full null space. However, it is not hard to imagine larger examples in which the number of null matrices in the full null space is unknown. This shows that if we used a purely uniform random construction method to produce, say, 10,000 sample null matrices (with duplicates because there are only five unique solutions), there would be approximately (2,060, 2,130, 2,199, 2,130, 1,481) of each—and this is not a uniform random distribution! A purely uniform random distribution would result if there were a more or less equal number of each member in the sample null space. Remember that this is a simple example and each member of the sample null space should be unique. The point of this example is that construction algorithms that choose unfilled indices uniform randomly do not produce a uniform random sample null space.

This is an extremely disappointing result, but in retrospect it is not surprising. The constraints act to direct the filling procedure. Moreover, a close inspection of the five members of the full null space shows that certain indices are filled more often than others. For instance, index $(1, 1)$ is filled in four of five members, while index $(3, 3)$ is filled only once in five. Because filled indices are not uniformly distributed in the full null space, an algorithm that chooses indices uniform randomly for filling is doomed to fail. In fact, a biased construction algorithm is needed. The bias, however, is not known a priori.

Can Swapping Methods Do Any Better?

For the same reasons as above, unfortunately the answer to the heading's question is no. The swapping algorithm suffers from the same problem. By applying more swaps, all solutions can be created but with different probabilities. When applied to **A**, one swap produces **C**, **D**, and **E**, with probability 5/16, 5/16, and 3/8, respectively. **B** cannot be created from **A** in one swap. Two swaps produce **A**–**E**, with probability 74/32, 24/32, 79/32, 24/32, and 55/32, respectively. Subsequent swaps did not equalize the probability of each member occurring. Swapping algorithms do not produce a uniform random sample null space.

In both examples, repeat solutions were not discarded. This was simply to demonstrate the nonuniform random nature of null matrix creation. In even slightly more complicated examples, the full null space will be unknown and the bias will go unnoticed.

The Trial-Swap Algorithm

Miklós and Podani (2004) overcame the difficulties we've described in their trial-swap algorithm. Miklós and Podani presented a proof that their algorithm produced a sample null space that was representative of the full null space. Because their algorithm was computationally inefficient for large problems, Miklós and Podani suggested an alternative algorithm that showed good agreement with known solved problems. We present their alternative algorithm here.

For sample null space creation, the trial-swap method of Miklós and Podani is fast and effective. The trial-swap algorithm begins as a construction algorithm and creates each sample null member from a matrix of zeros. The trial-swap algorithm also uses a swap algorithm to ensure that the sample null space is approximately uniform random.

Recall that the goal is to create a sample null space representative of the full null space. Each member of the full null space should (in theory) have an equal chance of representation in the sample null space. The trial-swap algorithm creates a sample null space one member at a time, starting from a zero matrix, $\mathbf{A}$, and the row and column sums derived from the observed incidence matrix.

Sample null matrix creation involves several steps. These steps are first summarized and then provided in more detail below.

1. Compute the row and column sums of the observed incidence matrix. Steps 2–10 are repeated to create each new member of the sample null space.
2. Randomize the order of the row sums and column sums.
3. Beginning with a zero matrix $\mathbf{A}$, fill $\mathbf{A}$ trivially so that the row and column constraints are conserved. There is no requirement that $\mathbf{A}$ be a binary matrix of zeros and ones.
4. Transform A into a binary matrix using quasi-swaps.
5. Apply K trial-swaps to $\mathbf{A}$ to create a member of the sample null space.

Details of steps 3–5 are given below using the Galápagos finches as an example.

Step 1 is trivial. The row and column sums are $\{(14,13,14,10,12,2,10,1,10,6,2,6,11,11), (4,4,11,10,10,8,9,10,8,9,3,10,4,7,9,3,3)\}$.

In step 2, without loss of generality, assume the randomized order is as given in step 1. There is no need, for example, to sort the row and column totals; any order will do.

Step 3 is satisfied by filling $\mathbf{A}$ to preserve the row and column sums using the following algorithm. The resulting $\mathbf{A}$ is not binary.

Let $\mathbf{A}_{1,1}$ equal the first row sum or first column sum, whichever is the least. If $r_1 = c_1$, then either can be chosen. A shorthand notation suffices:

$$\mathbf{A}_{1,1} = \min (r_1, c_1)$$

In our example, $\mathbf{A}_{1,1} = 4$ since the first column sum is less than the first row sum (14). (See table 5.1 below.) Subtract $\mathbf{A}_{1,1}$ from the first row sum r_1 to

88

obtain $r_1 = 10$, and first column sum c_1 to obtain $c_1 = 0$. Note that either the first row or first column constraint (or both) is (are) satisfied, meaning either $r_1 = 0$, or $c_1 = 0$, or $c_1 = r_1 = 0$. What remains of r_1 or c_1 must then be assigned.

If $c_1 = 0$, the first column of $\mathbf{A}$ is filled, otherwise $r_1 = 0$ and the first row of $\mathbf{A}$ is filled. If $c_1 = r_1$, then both the first row and first column of A are filled. Continuing,

$$\text{If } c_1 = 0, \text{then } \mathbf{A}_{1,2} = \min (r_1, c_2).$$

$$\text{If } r_1 = 0, \text{then } \mathbf{A}_{2,1} = \min (c_1, r_2).$$

$$\text{If } c_1 = r_1 = 0, \text{then } \mathbf{A}_{2,2} = \min (c_2, r_2).$$

In our example, $\mathbf{A}_{1,2} = \min (10,4) = 4$ (see table 5.1). Subtract $\mathbf{A}_{1,2}$ from c_2 so that $c_2 = 0$, and from r_1 giving $r_1 = 6$ (see table 5.1). In essence, what remains is what must be allocated in the next and subsequent steps. A similar step gives $\mathbf{A}_{1,3} = 4$, which completes the filling of the first row of $\mathbf{A}$. Note that the first two columns were also filled. In essence, the above steps first fill the smaller of what remains of the row and column sums.

Thus, $\mathbf{A}$ is systematically filled by row and by column according to the row and column sums. The filling procedure is guaranteed to be completed successfully. The result is that $\mathbf{A}$ is more or less a quasi-diagonal matrix whose row and column sums after randomization are identical to the observed incidence matrix (table 5.1).

In step 4, $\mathbf{A}$ is transformed into a binary matrix using quasi-swaps. A quasi-swap is performed on $2x2$ submatrices (whose terms need not be adjacent rows or columns but form a submatrix, meaning that (w,y) and (x,z) are in the same column; (w,x) and (y,z) are in the same row) of the form

$$\begin{matrix} w > 0 & x = 0 \\ & & \text{becomes} \\ y = 0 & z > 0 \end{matrix} \qquad \begin{matrix} w - 1 & 1 \\ \\ 1 & z - 1 \end{matrix}$$

so that the diagonal terms are reduced by 1 and the off-diagonal terms are increased by 1. In this way, a quasi-swap is a reduction step, reducing the diagonal terms by increasing the off-diagonal terms and thus preserving the row and column sums. For instance, the $3x3$ matrix below can be transformed into a binary matrix

TABLE 5.1 **Matrix A is filled so that the row and column sums are preserved (note that A is not a binary matrix)**

Row totals	4	6	11	13	10	12	9	13	8	12	3	13	4	11	10	7	3
14	4	6	4														
13			7	6													
14			7	7													
10				3	7												
12					5	7											
7						2	5										
10							8	2									
8								6	2								
10									10								
9										3	6						
8											7	1					
10												3	7				
13													4	9			
11														1	7	3	

$$
\begin{array}{ccc}
\mathbf{2} & \mathbf{0} & 0 \\
\mathbf{0} & \mathbf{1} & 0 \\
0 & 0 & 1
\end{array}
$$

with a single quasi-swap applied to the submatrix in bold as follows:

$$
\begin{array}{cc}
\mathbf{1} & \mathbf{1} \\
\mathbf{1} & \mathbf{0}
\end{array}
$$

that results in the binary matrix:

$$
\begin{array}{ccc}
1 & 1 & 0 \\
1 & 0 & 0 \\
0 & 0 & 1
\end{array}
$$

After a series of quasi-steps, the matrix for Galápagos finches begins to transform into a binary matrix (table 5.2).

With repeated quasi-swaps, A can be reduced to a binary matrix whose row and column sums match those of the observed incidence matrix. Note that in rare instances, it is possible after repeated quasi-steps to create a matrix with elements >1 but whose form no longer permits additional quasi-steps to be performed. In this case, the matrix can be rejected and the procedure restarted. During the creation of one million Galápagos finches sample null matrices, this event happened 12,695 times.

TABLE 5.2 **Repeated steps transform A to binary form**

Row totals	Column totals																
	4	6	11	13	10	12	9	13	8	12	3	13	4	11	10	7	3
14	1	1	1	1	1	1	1	1	1	1	1	1	1	1			
13	1		7	5													
14	1			7	6												
10	1				3	6											
12		1					5	6									
7		1					2	4									
10		1							8	1							
8		1							6	1							
10												10					
9		1	1								2	5					
8			1									7	0				
10			1										3	6			
13													4	9			
11															1	7	3

Step 4 completes when **A** is a binary matrix whose row and column sums match those of the observed. Step 5 involves applying K trial-swaps to **A**. K is given a priori as twice the number of 1's (presences). For each trial swap attempt an index, say (i,j), of **A** is chosen randomly. Another index, say (k,l) is also chosen randomly. If a valid swap can be performed with **A** (i,j), **A** (k,l), **A**(k,j), and **A** (i,l), then a swap is performed, otherwise no swap is performed. In any case a trial is counted. Thus, K trials are made to apply a simple swap to **A**, only some of which are successful. After K trials or swap attempts, **A** is added to the sample null space. According to Miklós and Podani (2004), this procedure produces a near-uniform random sample null space.

Summary

In this chapter, we present algorithms and examples to create the sample null space. Connor and Simberloff (1979) and others adopted the use of a swap algorithm to create new null space members because of the speed at which a swap algorithm could generate new members of the null space. Because the generation of new members starts with the observed community or the previous random community, there is always danger that the newest random member bears a strong resemblance to its predecessor. The number of swaps necessary to ensure that the new member is independent of its predecessor is unknown. Clearly the number increases as the number of rows and columns in the community increases, but we do not know how many are needed to create a truly independent member. That is why starting from a clean slate, a community of all absences (zeros), offers advantages.

Construction algorithms generate new members of the null space by starting from a matrix of 0's and filling it carefully so as not to violate a priori constraints. To guarantee that successive null matrix members were independent, each construction of a null member began with a matrix of 0's, a mathematician's null matrix.

Enumeration algorithms generate new sample null space members sequentially and cannot be terminated prematurely without loss of sampling part of the full null space. Because of this handicap, enumeration algorithms are used to generate the full null space but can only be used for small problems.

We showed that neither swap nor construction algorithms that choose indices equiprobably (with uniform probability) lead to the creation of a sample null space that is a perfectly uniform random sample of the full null space. This is because each index in the full null space is not filled (contains a presence) in the same proportion as all other indices.

The trial-swap algorithm of Miklós and Podani (2004) starts with a zero matrix and uses a construction algorithm and then a simple swapping algorithm to create a uniform random sample null space. Miklós and Podani provided a proof that the trial-swap algorithm produces a uniform random sample null space. However, because the algorithm was too slow in practice, they suggested a modified method and presented evidence that the method produces an approximately uniform random sample null space. The trial-swap method is the method of choice for creating sample null spaces.

CHAPTER SIX

How to Characterize
Incidence Matrices

If nonrandom distributions are detected by the analysis of Connor and Simberloff (1979), one only knows that a large number of species are interacting and not which species. — Wright and Biehl 1982

Then You Need a Metric . . .

Recall that the first step in the analysis of communities was to create a representative collection of random, or null, matrices (the sample null space) that had the same properties as the observed incidence matrix. The previous two chapters presented the difficulties of creating a representative sample of null matrices subject to ecologically sensible constraints and showed how these difficulties were overcome. The second essential step is to compare the sample null space to the observed incidence matrix. The obvious question is: how? Comparison is all a matter of the metric one uses. As will become apparent, choosing the metric is a bit like choosing a jury—choose the right one and you can get the desired result.

In previous chapters, we have often used quite simple metrics such as whether two species occupy mutually exclusive sets of islands, or perhaps co-occur on one or a few islands. The obvious problem is where to draw the line. Just how many are more than "a few." So, in this chapter, we examine the history of the various choices of metrics.

We first discuss *ensemble* metrics. Ensemble metrics are generally constructed by summing a measure, for example, the number of co-occurrences of a species pair, across all possible species pairs, and then dividing by the total number of pairs to give an average value of the measure. In the case of Vanuatu, for instance, this means summing up the number of co-occurrences of all 1,540 species pairs and dividing the total by 1,540. The so-called numbers-of-pairs metric and the single number it yields are presumably like the speed of light, c, or the mathematicians' π, or Euler's e. We find, however, that such a metric effectively obscures all unusual patterns within an average. For this reason, we refer to ensemble metrics as *cloaking metrics*.

The Metric of Connor and Simberloff (1979)

In their initial response to Diamond (1975), Connor and Simberloff had to invent a measure, or metric, by which to compare patterns in the observed incidence matrix to those same patterns derived from the collection of null matrices, the sample null space. There was no popular metric so the choice was entirely up to Connor and Simberloff: "We then scanned each simulated arrangement for the numbers-of-pairs not found anywhere (and number of trios for New Hebrides birds and West Indies bats), numbers-of-pairs (or trios) found on only one island, only two islands, only three islands, etc. Finally, we examined the actual arrangements. All analyses were performed with and without constraint (iii) 'incidence functions.' However, since relaxing incidence constraints does not affect the results, only the results including incidence constraints are presented" (1979, 1133).

For their analysis, Connor and Simberloff (1979) chose as a metric the number of co-occurring pairs of birds on islands. That is, they computed the numbers of pairs that did not co-occur (checkerboards), the numbers of pairs that occurred once-and-only-once, twice-and-only-twice, . . . 27-and-only-27 times, and 28-and-only-28 times (there were 28 islands in the Vanuatu archipelago).

When one plots the numbers of such pairs derived from the average of ten null matrices against that derived from the observed incidence matrix, the agreement is striking (fig. 6.1). The difference between the line and the points is often so small that one cannot distinguish them. Connor and Simberloff (1979) then used a χ^2 goodness-of-fit test to compare the differences between the numbers of pairs in the observed incidence matrix to the numbers of pairs derived from the sample null space. They demonstrated statistically the close agreement that is obvious in the figure (Connor and Simberloff 1979, 1134).

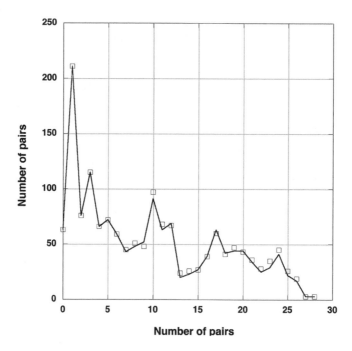

FIG. 6.1. The numbers of species pairs that co-occur on no islands (so forming checkerboards), on one island, two islands, etc., for land birds of the New Hebrides (now Vanuatu). The dots indicate the observed values; the line is that produced by Connor and Simberloff's null model.

Even the most casual reader could not help but see clearly that the observed numbers of co-occurrences did not differ from random expectations. The fit was striking, the implication clear, and the evidence against Diamond irrefutable. The jury of peers could reach only one outcome: assembly rules were a figment of Diamond's imagination.

The close fit is just as it should have been. As we showed in the previous chapter, the swapping algorithm Connor and Simberloff employed, plus the fact that they only used a sample of ten null matrices, meant that those nulls would inevitably be very similar not just to each other but to the observed incidence matrix as well. In this chapter, we provide another reason: the similarity is a consequence of the metric used.

Fully three years elapsed before Diamond responded. Diamond sought the help of his colleague Michael Gilpin at the University of California, San Diego. Of all the possible loopholes in the work of Connor and Simberloff (1979) that could have been exploited, all that Diamond and Gilpin (1982), and later Roberts and Stone (1990), could come up with was to criticize the use of the χ^2 goodness-of-fit test on the basis that the data used in the test were mutually dependent. Of course, this was true, but it was a weak rejoinder. The burden of proof was on Diamond to argue that his results were unusual, not that his critics had made a statistical error.

Connor and Simberloff's metric was computed by summing all possible species pairs—a cloaking metric—that very effectively obscured subtle patterns within a mass of irrelevant noninteracting species pairs. Connor and Simberloff found that of all 1,540 possible pairs, 63 species pairs in the observed incidence matrix failed to co-occur on any islands—they formed a checkerboard. In the sample null space, they found that, on average, 63.2 species pairs failed to co-occur. The standard deviation was presumably small, and so they concluded that this was not in any way unusual. The choice of metric obscured those few species pairs that might have had an unusual checkerboard, that is, ones that differed significantly from chance expectations.

In chapter 3 we discussed the fact that most of the 63 pairs are the consequence of just a few species that live on just one island and hence form checkerboards with species that live only elsewhere. What the ensemble metric does is take all the possible pairs and lump them together: those few that are in fact unusual are lumped in with the vast majority that are not. An analogy serves to illustrate this critical point.

Suppose we have a can of Pringles Potato Chips®, and we want to know if any of the chips in the can differ significantly in weight from other chips. What Connor and Simberloff would have us do is weigh our can and then compare the weight to the average weight of ten cans. Note that the few possibly unusual chips are but a perturbation on the total weight in the can; they have been lumped together with the vast majority of chips that do not differ in any way from the average. Though our can of Pringles might contain some lightweight chips, we would not know unless many were underweight. The metric chosen by Connor and Simberloff (1979) is the perfect metric by which to obscure patterns, just as Harvey et al. (1983) stated. With this metric, coupled with the sample null space, it would have been miraculous if any patterns were uncovered.

Second, of what possible ecological significance is the co-occurrence of two completely unrelated or noninteracting species? For instance, is it important that a duck and a flycatcher might never co-occur? Should we not expect that of 1,540 possible species pairs, many would likely never co-occur just by chance alone—just as Connor and Simberloff (1979) argued to demolish assembly rule 1 (rule a in their notation)? On the one hand,

Connor and Simberloff (1979) argued logically that it was to be expected that of all possible pairs of birds, some would never co-occur. But is it unusual that two Galápagos finches—each found on half the Galápagos Islands, both in the same genus, both weighing about the same, both consuming about the same food, and both nesting at the same time—never co-occur on any islands? We cannot know using the numbers-of-pairs metric.

The second numbers-of-pairs value was the number of species pairs that co-occurred once-and-only-once. That is, we compute the total number of all species pairs that occurred once-and-only-once. If, for instance, such a species occurs on the most species-rich island, then its contribution to the once-and-only-once number-of-pairs can be significant. If, on the other hand, such a species was one of Diamond's supertramps that occurred only on a single species-poor island, then its contribution to the once-and-only-once sum was small. Again, there was an intimate connection between variability in the sample null space and the metric.

Not surprisingly, Connor and Simberloff (1979) again found no significant difference between the number of pairs of species that co-occurred once-and-only-once in the observed incidence matrix and the same number derived from the sample null space. This was because, upon transposition, the column with a single-island species can interchange a 0 with a 1, or vice versa. Certainly, changes occurred as elements of the observed matrix were interchanged, but these changes amounted to small perturbations. In any case, the procedure was performed very few times. Each step perturbed the numbers of occurring/nonoccurring pairs only slightly from the observed incidence matrix and, as noted by Diamond and Gilpin (1982), was not random. When this analysis was repeated using a larger and more random sample null space, Sanderson et al. (1998) found there was a statistically significant difference in the metric values. The difference can be attributed to the rich sample null space created by a construction method. This was indeed a fortuitous event because the ensemble numbers-of-pairs metric is also useless in uncovering unusual patterns.

Wright and Biehl (1982)

Wright and Biehl (1982) argued against the use of the numbers-of-pairs metric as well as the null model Connor and Simberloff (1979) employed. They presented an alternative method of analysis based upon the use of a hypergeometric function that we discussed in the previous chapter. Wright and Biehl (1982) also raised another issue of great import—one that Diamond (1975) recognized repeatedly in his original analysis, and one completely ignored by Connor and Simberloff (1979). Wright and Biehl wrote: "The problem is exacerbated if the species to be analyzed are delimited taxonomically and without regard to autecologies. For example, the class Aves includes species as disparate in their ecologies as humming-birds and vultures. These species do not interact, and they should not affect one another's island distributions" (1982, 347).

This is why Diamond (1975) repeatedly analyzed guilds. Simberloff's comeback was that the use of guilds in analysis required that every species be assigned to a guild, and this was impossible. But Simberloff's rejoinder, too, is incorrect. Why not simply analyze species pairs just as Wright and Biehl (1982) had suggested, or perhaps a single guild? This concern was apparently ignored until Sanderson (2000) used this metric to analyze the Galápagos finches one pair at a time. In subsequent chapters we specifically look at species belonging to the same genus—something one can assign a priori.

In 1983, it was time for a synthesis of the use of null models in ecology. One was provided in the *Annual Review of Ecology and Systematics* (Harvey et al. 1983).

Harvey et al.'s (1983) Review of Null Models in Ecology

Null models analyze various problems in ecology: the relative abundance of species, the number of species, species co-occurrence patterns on archipelagos and gradients, and other occurrences. Harvey et al. (1983) provided a thoughtful summary of progress and problems. They wrote: "The islands of an archipelago constitute a series of alternative microcosms. Communities differ somewhat among islands and, within a particular guild or taxonomic group, the presence of some species may be associated with the presence or absence of others. Detecting such patterns provides a promising start toward analyzing structure in natural communities" (194).

They discussed the need to constrain the null space by using the row and column constraints. As for the incidence constraint, they wrote: "a major problem with incorporating the constraints is that relative species abundance and incidence functions may well have been shaped by competition" (197). While this might be true, absolute or relative abundances were not included in the incidence matrix, and the use of the incidence constraint is not widespread. Species abundances have been used to support conclusions derived from the analysis of species co-occurrences.

Harvey et al. (1983) were clear regarding ensemble metrics: one should avoid them. They specifically referred to the statistic of Connor and Simberloff (1979) that is referenced (20) in their paper: "The problem of high type II statistical error may be exacerbated by the use of statistics that have weak discriminatory ability. For example, one test (20) examines all pairwise species combinations and counts how many of them are found on various numbers of islands—33 species pairs on no islands, 14 species pairs on one island and so on."

Note the coining of the phrase *weak discriminatory ability*. Connor and Simberloff (1979) exercised strong discriminatory ability by selecting a metric with exceptionally *weak discriminatory ability*. Diamond and Gilpin had earlier written: "this test greatly understates the case [because] buried in this total are pairs of ecologically close species with exclusive distributions despite each species occupying many Bismarck islands, such that the probability of attaining such a result by chance is as low as 6×10^{-9}" (1982, 71).

This statement is certainly true. Typically, one performs tests for significance at the 95% confidence level. What this number implies is that a priori the vast majority of patterns will not be found to differ from chance expectations. This being the case, why would one use an ensemble metric at all? Unfortunately, most subsequent researchers such as Stone and Roberts (1992, 1990), Roberts and Stone (1990), Gotelli (2001), his mentor Simberloff, and his colleagues completely ignored such admonitions.

Harvey et al. concluded that "the data in the species × island occurrence table is insufficient to enable one to distinguish among the possible explanations. Of course, the additional data only provide circumstantial evidence that make particular explanations more or less likely to be correct. At the very least, the consequences of introduction and removal experiments can now be predicted with more certainty" (1983, 198).

Yes! What this says is that structuring mechanisms cannot be inferred from the observed incidence matrix. One can subject an unusual pattern to further research (and sometimes experiments at small scales), however, to help reveal structuring mechanisms. Later we give an example from the literature regarding a removal experiment of frogs inhabiting a stream (Inger and Colwell 1977). Because we have learned so much about Galápagos finches, we can do further analyses to explain unusual co-occurrences patterns (chap. 7). First, we must identify unusual co-occurrence patterns.

Harvey et al. concluded that "the most complete analysis reported so far is Gilpin and Diamond's (1982) study on the birds of the Bismarck islands" (1983, 199). Later in 1983, Roughgarden weighed in. In a lengthy paper titled "Competition and Theory in Community Ecology," Roughgarden concluded: "This essay was provoked by the insistent claims of Connell (1980), Connor and Simberloff (1979), and Strong et al. (1979) that competition and the coevolution of competitors are not real and important processes in nature, and hence, that theory for these processes is not worthy of testing. These authors are wrong" (1983, 599).

Connor and Simberloff (1979) did not reject competition as a structuring mechanism. What they were saying is that we first must show that the observed pattern is unusual. If it is not, the principle of parsimony says that no simple structuring process is needed to explain the observed pattern. Alternatively, if we find the pattern to be unusual then we should investigate the structuring mechanisms. This sets a high standard of proof—"beyond reasonable doubt," as it were, rather than the "preponderance of the evidence."

In 1984, the book *Ecological Communities*, edited by Strong, Simberloff, Abele, and Thistle, contained a chapter titled "Rejoinders," one of the more memorable chapters. In this same volume Colwell and Winkler (1984) wrote about the Narcissus effect, whereby when one samples from a community that competition has already shaped, one will underestimate its effects because they are already reflected in the pool of species present.

The imposition of more constraints reduces variability in the sample null space, increasing the likelihood that a pattern in the observed community will not differ from chance expectations. Fewer or no constraints increase variability in the sample null space, thus increasing the likelihood that an observed pattern does differ from chance expectations. Those species, especially congeneric species, whose observed patterns differ from chance expectations might well offer clues to structuring mechanisms and so should be investigated.

Specifically, Colwell and Winkler realized that, for example, were one to require incidence constraints, one might well not infer competition, even if it were a major factor in shaping the community. Ecologically similar species might not overlap in their incidences—one occurring on small islands, another on larger ones. Constraining null models to have species occur on the ranges of islands sizes where they are found would mean that one would not find these patterns of mutual exclusion to ever be unusual.

One can clearly set up the standard of "beyond reasonable doubt" to be impossible to meet.

Colwell and Winkler also provided a valuable summary for the way null models should be designed and utilized (which repeated Diamond 1975):

> If interspecific competition occurs in a community it presumably will be strongest and hence most detectable in a small subset of ecologically similar species. Including other less similar species in an analysis runs the risk of obscuring the effects of competition. Community-level analyses are of great heuristic value in detecting community patterns. The patterns thus detected, however, must not be overinterpreted as proof of process. The assessment of any interspecific process such as competition will be most profitably pursued and the results most firmly established by an in-depth analysis of small groups of ecologically similar species identified in the community-level analysis. (1984, 359)

Much of this was subsequently ignored. Instead, the use of cloaking ensemble metrics that obscure ecological patterns was about to get a boost.

Stone and Roberts (1990, 1992) and Roberts and Stone (1990)

Roberts and Stone (1990) returned to the issue of the proper metric for which to analyze species on islands. Unlike Wright and Biehl (1982), and much to their credit, Roberts and Stone used a null model with both row and column constraints (i–ii). Like Wright and Biehl, Roberts and Stone also suggested comparing pairs of species. They argued that the number of islands shared should be the metric by which to test the null hypothesis. Unfortunately, they did not stop there.

Roberts and Stone (1990) defined S_{ij} as the number of islands shared by species i and j. We henceforth refer to this metric as the *natural metric* because it seems to us a very natural number to compute and use as a metric. They defined a sharing matrix **S** whose (i, j) entry was the number of island shared, S_{ij}. Clearly, a diagonal element, S_{ii}, is just the number of islands occupied by species i. Ignoring the admonitions of Wright and Biehl (1982) and Harvey et al. (1983), Roberts and Stone created a single number that would be the metric by which to test the entire community ensemble.

$$S^2 = \frac{1}{n*(n-1)} \sum_{i=1}^{n} \sum_{j-1, j \neq i}^{n} S_{i,j}^2$$

According to Stone and Roberts (1990), an unusually large C-score occurs when checkerboard patterns, that is, patterns of mutual exclusivity, predominate in a community, suggesting evidence for pairwise competitive exclusion. Conversely, a small C-score indicates many pairwise co-occurrences, suggesting other forces are more important. Why is it that the full richness of nature embodied in more than a billion years of evolution, the wondrous complexity and interactions of living organisms, must always be boiled, distilled, spun, and precipitated into a single number? That ecologists would place great value in such a number is in itself wondrous in its own mysterious simplicity.

To create the sample null space for Vanuatu, Roberts and Stone (1990) followed *the displace and perturb procedure* used by Connor and Simberloff (1979) but used grander perturbations. First, starting with the observed incidence matrix, **A**, they performed 100,000 swaps "to guard against the retention of any unusual qualities from **A**" (Roberts and Stone 1990, 563). Later, Gotelli (2000) would give a different reason for doing the same thing. Gotelli (2000) used 10,000 initial swaps to the original data set "to remove transient effects" and "retained the next set of 1,000 swaps to form the null set" (Gotelli and Entsminger 2003, 533). Each successive null space member differed from the previous member by a single swap. Later, Gotelli and Entsminger would criticize other methods for producing *large standard deviations*.

When Roberts and Stone (1990) applied the null model with the ensemble S^2 metric to the Vanuatu avifauna, they found that S^2 for the observed incidence matrix was 148.95, the average S^2 for the sample null space was 147.10, and the standard deviation, $\sigma = 0.23$.

Roberts and Stone could hardly contain themselves: "Thus the observed value of S^2 differs from the random-sample mean by 1.75, or a little over 1%. But any idea that the null hypothesis can therefore be accepted is quickly dispelled, when we note that this difference—tiny though it may appear—is 7.6 times the standard deviation (both mean and s.d. being estimated from the sample)" (1990, 564). They concluded that "the species distribution in the archipelago cannot plausibly be regarded as arising only from the process implied by the null hypothesis."

They wrote, however, that in the sample null space the maximum value of $S^2 = 147.79$. We found this value to be curious. Recall that the Vanuatu avifauna consists of 56 birds on 28 islands and the observed incidence matrix is half full. Thus, variation in the full null space should be large, and the maximum value of any ensemble metric should differ more than 1% from the average value. In our reanalysis (Sanderson et al. 1998) of the Vanuatu avifauna using the metric of Roberts and Stone (1990), we used a sample null space of 5,000 matrices created by a construction algorithm. We found for the sample null space $S^2 = 150.47$ (max = 203.651) with $\sigma = 1.242$. We concluded that the distribution did not differ from a random expectation—*using the S^2 metric*. When this conclusion was published, Stone (pers. comm.) asked to see some of the sample null matrices with S^2 in the range of 200. By then Sanderson had created a sample null space of some 100,000 unique null matrices. Sanderson e-mailed Stone several hundred matrices and their S^2 values!

What is going on? Here is what we think happened to Roberts and Stone (1990). The original swap perturbation created a matrix with little variation—and this is a characteristic of swap methods applied solely to the observed incidence matrix. As we showed in the previous chapter, some sample null matrices are more equal than others. They act as attractants, creating a vortex into which sample matrices fall and are unable to escape. We call these null matrices *swap black holes*. The subsequent perturbations that created the sample null space all ended up sampling in the *neighborhood* of the perturbed sample null matrix. This explains the small standard deviation and the small maximum S^2 being so close to the average value of the metric. Later, Gotelli and Entsminger (2001) criticized the knight's tour by suggesting that the method led to excessively large standard deviations. Gotelli and Entsminger (2001) made no attempt to compare the sample null spaces created by different methods to a known full null space.

A few months later Stone and Roberts (1990) published the checkerboard score, or C-score, metric. Once again rejecting the advice of Wright and Biehl (1982), Harvey et al. (1983), and others that ensemble metrics do not work, ignoring Diamond's original analysis of guilds, and wrapping up everything into a single magic number, Stone and Roberts brought forth yet another cloaking metric that again took into account every possible species pair. This time, however, Stone and Roberts (1992) targeted checkerboards—or so it appeared.

Why Checkerboards?

Diamond (1975) originally suggested that since competition for resources had structured certain guilds, the most obvious evidence would be patterns of mutual exclusivity or checkerboard patterns. Diamond wrote: "The simplest distributional pattern that might be sought as possible evidence for competitive exclusion is a checkerboard distribution. In such a pattern, two or more ecologically similar species have mutually exclusive but interdigitating distributions in an archipelago, each island supporting only one species. . . . Checkerboard distributions are of great interest in demonstrating the existence of competitive exclusion" (1975, 387–88, 392).

Diamond and Gilpin reiterated this belief, stating that the "simplest and clearest pattern that might be produced by competition is a checkerboard distribution" (1982, 65). Stone and Roberts incorporated this notion into their C-score metric. Their reasoning was that unusual communities would have more checkerboards that those that did not differ from chance expectations. Using their previous recent work, they defined S_{ij} to be the number of co-occurrences of species i and j, and let r_i be the ith row sum, that is, the number of times species i occurred. Then

$$C_{ij} = (r_i - S_{ij}) \times (r_j - S_{ij})$$

is a number that is maximized when species i and j each occurred on half the islands but did not co-occur on any islands. In any case, for m = number of species in the community, the C-score is given as

$$\text{C-score} = \frac{2}{n * (n-1)} \sum_{i=1}^{n-1} \sum_{j=i+1}^{n} C_{i,j}$$

where n is the number of islands.

The leading term is the reciprocal of the number of all possible unique pairs of species. Once again, the sum over all possible pairs would obscure those few that might be unusual. Nonetheless, Stone and Roberts (1990) turned to the Vanuatu avifauna. The observed incidence matrix had a C-score = 9.53. For a sample null space created by using swaps, they found C-score = 9.128, σ = 0.072. Similarly, for a second sample null space created by a construction method they found C-score = 9.112, σ = 0.067. In both instances, they found that the C-score of the observed incidence was unusual.

Unlike Connor and Simberloff (1979), Stone and Roberts (1990) reported that the West Indies bats had a C-score that differed significantly from chance expectations. What does this mean? It means only that this particular metric, the average of all possible pair-wise C-scores, yielded a different conclusion from that obtained by Connor and Simberloff (1979). Just as Wright and Biehl (1982) suggested, this has absolutely no ecological significance—but then neither did the metric of Connor and Simberloff (1979). Why not simply test each C_{ij} separately? This would have to wait. The last paragraph of Stone and Roberts (1990) read: "Can we therefore conclude that evidence has been found here, for biological factors which tend to keep species apart? After examining this question, and probing more deeply with the C-score technique, we believe that the truth is somewhat more complicated than this; a report is in preparation."

The report issued in 1992 did not shake the foundations of community ecology. Two nuggets, however, were buried deep within it. Roberts and Stone introduced two ensemble metrics, the S and C-score, and followed these with a third ensemble metric, the *togetherness score*, denoted simply as T (Stone and Roberts 1992). Since we do not intend to dwell upon this further, there is no need to mathematically describe precisely how T is computed. Of course, the three metrics were related (Stone and Roberts 1992).

Why Ensemble Metrics Fail—An Example

Gotelli (2000) recommended the C-score as a metric that could detect patterns of co-occurrence consistent with competitive exclusion. We now test these beliefs with an example.

One of the foundations of mathematics is that while it is impossible to *prove* a theory is correct by providing countless examples of its truth, a single counterexample can *disprove* the theory. Consider a simple example of 15 species occupying ten sites (table 6.1).

The unnormalized C-score of the observed incidence matrix is 324 (the sum was not normalized by the total number of species pairs 15(14/2) to facilitate comparisons). Note that in this example, there are many species pairs that show checkerboard patterns, that is, they do not co-occur. For instance, species pairs (1, 2), species pairs (3, 4), species pairs (5, 6), and all possible species pairs (7–15) form checkerboards. Other species pairs such as (1, 12–15), (2, 7–11), (3, 8), (3, 10), (3, 12), and (3, 14) and others also do not co-occur.

TABLE 6.1 **An example of 15 species that occupy 10 sites**

	Sites									
Species	1	2	3	4	5	6	7	8	9	10
1	1	1	1	1	1	0	0	0	0	0
2	0	0	0	0	0	1	1	1	1	1
3	1	0	1	0	1	0	1	0	1	0
4	0	1	0	1	0	1	0	1	0	1
5	0	0	1	1	1	1	1	0	0	0
6	1	1	0	0	0	0	0	1	1	1
7	1	0	0	0	0	0	0	0	0	0
8	0	1	0	0	0	0	0	0	0	0
9	0	0	1	0	0	0	0	0	0	0
10	0	0	0	1	0	0	0	0	0	0
11	0	0	0	0	1	0	0	0	0	0
12	0	0	0	0	0	1	0	0	0	0
13	0	0	0	0	0	0	1	0	0	0
14	0	0	0	0	0	0	0	1	0	0
15	0	0	0	0	0	0	0	0	1	0

Note. There are large numbers of checkerboards where two species do not co-occur.

Stone and Roberts (1990) want to know if the observed C-score differs from chance expectation. They could also have asked a more direct question: does the average number of observed checkerboards differ from chance expectations? Likewise, they could have asked if the total number of observed checkerboards differed from chance expectations since the average is the total divided by the number of possible pairs. The answer to the basic question of whether either of these ensemble metrics was in any way relevant to the study of ecological communities was, we suppose, self-evident at least to these authors.

Using the methodology of Stone and Roberts (1990), we answered the first question by computing the C-score for the observed community and applying a Student's t-test comparing the observed C-score to the C-score and its standard deviation derived from the sample null space. Of course, this works when the distribution of C-scores and checkerboards is normal and should not be applied if the distribution is anything other than normal. The test to see whether the C-score is derived from a sample null space that is normally distributed is rarely done. A simple graph of the C-scores could be used in lieu of a formal statistical test (fig. 6.2).

For the *15x10* example (table 6.1), the observed unnormalized C-score and average number of checkerboards was 324 and 66, respectively. For a sample null space of 100,000 unique matrices, the average unnormalized C-score was 323.849 and average number of checkerboards was 66.35.

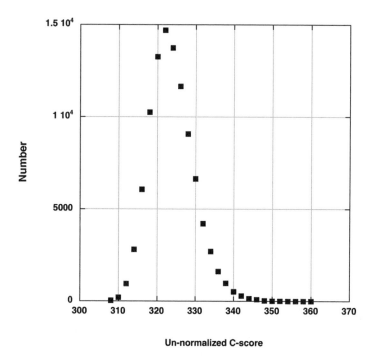

Un-normalized C-score

FIG. 6.2. The distribution of the unnormalized C-score obtained from a sample null space of 100,000 unique random communities. The observed unnormalized C-score was 324 and is not unusual.

Their standard deviations were 5.763 and 1.328, respectively, showing that neither the observed unnormalized C-score nor the average number of checkerboards differed from chance expectations. By the logic of Connor and Simberloff (1979), Stone and Roberts (1990), Gotelli (2000) and others, we are lead to conclude that patterns in *the observed community are most parsimoniously explained by chance.* Apparently, there are a substantial number of members of the sample null space that have greater values of C-scores and greater numbers of checkerboards (fig. 6.2).

Stone and Roberts (1990) asserted that the C-score is maximized when the observed community has a large number of checkerboards. We tested this claim directly using the sample null space of 100,000 unique members that satisfied the row and column constraints (i–ii). Table 6.2 supports their assertion.

Since the C-score and the number of checkerboards are proportional and because Diamond (1975) suggested checkerboards were the clearest

TABLE 6.2 **A comparison of the C-score and the number of checkerboards**

C-score	No. of C-scores	Average no. of checkerboards
308	26	64.00
310	219	64.01
312	860	64.57
314	2,787	64.85
316	6,058	65.14
318	10,113	65.48
320	13,294	65.75
322	14,789	66.05
324*	13,452	66.36
326	11,917	66.68
328	9,056	67.00
330	6,567	67.29
332	4,275	67.63
334	2,808	67.85
336	1,588	68.21
338	999	68.42
340	541	68.82
342	324	69.07
344	137	69.23
346	99	69.46
348	51	69.04
350	14	69.43
352	15	69.67
354	4	70.50
356	4	69.00
358	1	69.00
366	1	70.00

Notes. For the *15x10* example shown in table 6.1, we generated 100,000 unique sample null communities (the sample null space). The asterisk (*) indicates the unnormalized C-score of the observed community, which has many checkerboards, by design. It is not, however, unusual: the unnormalized C-score in the sample null space member with the greatest average number of checkerboards was 366.

indication of competition, one might conclude that the community was not structured by competition or, at the very least, that we cannot tell the difference between chance and competition in this community.

Intriguing questions remain unanswered, however. Simply because the C-score of the entire community does not differ from chance, we cannot conclude that each and every species pair does not differ from chance expectations. Inspection of the observed *15x10* incidence matrix reveals that six species each occur five times and form pairwise checkerboards: (1, 2), (3, 4), and (5, 6). Note also that the endemic species (those that occur once) also form checkerboards. Has the ensemble metric obscured interesting species co-occurrences by burying them in a *mass of distributional data*? The answer is yes; how could it be otherwise? To see this clearly, one must examine each species pair.

To be useful, a null model analysis must reveal those species pairs that differ from chance expectations that are embedded within a community. This suggests a metric based solely on the pair under investigation. We prefer the natural metric defined as the number of co-occurrences of a species pair, that is, S_{ij} of Roberts and Stone (1990).

A null model analysis using the natural metric can thus target a specific species pair by comparing the observed number of co-occurrences (a single value) to the number of co-occurrences distribution in the sample null space, that is, the collection of random communities. We need not rely upon the result being a normal distribution either. The sample null space, in statistical parlance, gives rise to a *distribution-free probability density function* of C-scores. The C-score of each member of the sample null space can be calculated, and each C-score can be counted. Only so many C-scores are possible in the sample null space, and many will be repeated. The distribution is the frequency of occurrence of each C-score in the sample null space. The probability density function is distribution free because it assumes no particular familiar form (such as normal, for example). It is a probability density function because the sum of the frequencies of the C-scores divided by the size of the sample null space is unity.

The distribution of the unnormalized C-score can thus be used to determine if the observed unnormalized C-score is in any way unusual. Because the C-scores form a distribution-free probability density function, we need only ask if the observed C-score is found in one or the other tail of the distribution in the null space. A Student's t-test or any other statistical test need not be employed. For the *15x10* observed incidence matrix, the observed C-score was found in 13,452 members of the sample null space and was therefore not in any way unusual. The observed average number of checkerboards—66—was not unusual (table 6.2). This avoids the problem of whether or not the distribution of C-scores derived from the sample null space is normal. The same can be said for the average number of checkerboards metric and for other metrics.

In the sample null space of 100,000 unique members, the unnormalized C-score varied from 308 found in 26 members of the sample null space to 366 found in just one member of the sample null space (table 6.2). The distribution of the C-scores derived from the sample null space could be closer to normal, but any differences will not likely change the result. We know this is true because we compared the observed C-score to the distribution-free probability density function (fig. 6.1).

We can use the above *15x10* community to demonstrate the effectiveness of the natural metric. The question is this: Which species pairs cooccur more or less often than chance expectations? Recall that Colwell and Winkler (1984) suggested that since patterns in ecological communities were created by some mechanism, it would be impossible to deduce whether a pattern differed from chance expectations (the Narcissus effect). Connor and Simberloff (1979) believed patterns that differed from chance expectations could be identified with a null model.

The basic form of the null model we now employ is that by Connor and Simberloff (1979). The first step is to create a sample null space whose members, all unique, are representative of the observed community. Second, the natural metric, the number of co-occurrences, is computed for each species pair in the observed community. Each species pair will be tested separately. Third, for each and every species pair, the observed number of co-occurrences, the natural metric, is compared to the frequency distribution derived from the sample null space. Those numbers of co-occurrences that are found to number less than 5% in the sample null space are said to be unusual (or differ significantly from chance expectations).

For instance, in the *15x10* example, species 1 and species 2 each occur on five islands, but not together—they form a checkerboard. In the sample null space of 100,000 unique members, species 1 and species 2 are found to form a checkerboard 1,123 times out of 100,000, co-occur once-and-only-once 1,720 times, twice-and-only-twice 45,815 times, three-and-only-three 30,127 times, four-and-only-four 5,094 times, and five-and-only-five just 121 times. Thus, the observed co-occurrence pattern of species 1 and species 2 is found in approximately 1.1% of the members in the sample null space, and so species 1 and 2 co-occur significantly less often than chance expectations (table 6.3). Similarly, species pairs (3, 4) and (5, 6) also co-occur significantly less often than chance expectations (table 6.3).

The possibility exists that two species might co-occur more often than predicted by chance alone. This example, however, does not have any such species pairs.

Note also that the checkerboards formed by the endemic species in the *15x10* observed community are not in any way unusual. For instance, species 1 and species 12 do not co-occur. In the sample null space of 100,000 unique members, however, species 1 and 12 form a checkerboard nearly 61,000 times and co-occur once approximately 39,000 times. Note also that species 2 and species 3 each occur five times and are observed to co-occur two times. Had these two species co-occurred five times in the observed incidence matrix, this would have been significantly unusual because it happened only in 123 of the 100,000 sample null space members. The C-score could never reveal these patterns.

Note that the resolving power of the natural metric to elucidate those species pairs that differ from chance expectations, either more or less, is something an ensemble metric such as the C-score could never achieve (tables 6.2 and 6.3). Moreover, the natural metric has immediate ecological appeal. Species pairs that co-occur significantly less often than expected *might* be exhibiting competitive exclusion, whereas those that co-occur more often than chance expectations *might* vary in body size, for instance, and consume different food. Those species pairs that differ significantly from chance expectations beg further investigation to determine why this is so. Much will depend on the ecology of the three unusual pairs in table 6.3. Is this not what ecologists really want to know?

Summary

In the study of ecological communities, we use a metric to test patterns observed in nature against the same patterns derived from the sample null space. The objective of the test is to determine whether the observed pattern differs from chance expectation as determined by comparing it to the sample null space. We must carefully consider the metric because it can obscure as well as enhance subtle patterns.

Diamond (1975) suggested that some patterns observed in the birds of Bismarck and Solomon Archipelagos were the result of competitive exclusion. He suggested that within some genera and guilds, the avifauna of these islands exhibited checkerboard patterns, meaning mutually exclusive distributional patterns.

In their effort to demonstrate statistically that Diamond (1975) was incorrect, Connor and Simberloff (1979) suggested that one should first test the patterns observed in nature by Diamond against a random, or null, community to determine whether such patterns differed from what would be expected were birds placed randomly on the islands.

When Connor and Simberloff (1979) used a numbers-of-pairs metric, they showed that patterns similar to those observed by Diamond in several well-documented communities such as the avifauna of the New Hebrides archipelago (now Vanuatu) and the bats of the West Indies did

TABLE 6.3 **All possible species pairs**

Species pairs		Observed	#Co	Sample null space number of co-occurrences					
				0	1	2	3	4	5
1*	2	5, 5	0	1123	17720	45815	30127	5094	121
1	3	5, 5	3	1123	17654	45473	30260	5339	151
1	4	5, 5	2	1113	17738	45760	30048	5183	158
1	5	5, 5	3	1117	17891	45383	30303	5150	156
1	6	5, 5	2	1169	17609	45377	30461	5241	143
1	7	5, 1	1	60333	39667	0	0	0	0
1	8	5, 1	1	60365	39635	0	0	0	0
1	9	5, 1	1	60379	39621	0	0	0	0
1	10	5, 1	1	60532	39468	0	0	0	0
1	11	5, 1	1	60433	39567	0	0	0	0
1	12	5, 1	0	60694	39306	0	0	0	0
1	13	5, 1	0	60118	39882	0	0	0	0
1	14	5, 1	0	60156	39844	0	0	0	0
1	15	5, 1	0	60406	39594	0	0	0	0
2	3	5, 5	2	1087	17735	45508	30412	5135	123
2	4	5, 5	3	1152	17674	45436	30439	5149	150
2	5	5, 5	2	1128	17687	45631	30263	5157	134
2	6	5, 5	3	1140	17384	45699	30550	5053	174
2	7	5, 1	0	60175	39825	0	0	0	0
2	8	5, 1	0	60516	39484	0	0	0	0
2	9	5, 1	0	60578	39422	0	0	0	0
2	10	5, 1	0	60320	39680	0	0	0	0
2	11	5, 1	0	60105	39895	0	0	0	0
2	12	5, 1	1	60351	39649	0	0	0	0
2	13	5, 1	1	60468	39532	0	0	0	0
2	14	5, 1	1	60472	39528	0	0	0	0
2	15	5, 1	1	60635	39365	0	0	0	0
3*	4	5, 5	0	1125	17925	45219	30323	5272	136
3	5	5, 5	3	1124	17791	45629	30104	5183	169
3	6	5, 5	2	1094	17666	45730	30105	5254	151
3	7	5, 1	1	60452	39548	0	0	0	0
3	8	5, 1	0	60515	39485	0	0	0	0
3	9	5, 1	1	60486	39514	0	0	0	0
3	10	5, 1	0	60513	39487	0	0	0	0
3	11	5, 1	1	60531	39469	0	0	0	0
3	12	5, 1	0	60366	39634	0	0	0	0
3	13	5, 1	1	60483	39517	0	0	0	0
3	14	5, 1	0	60177	39823	0	0	0	0
3	15	5, 1	1	60483	39517	0	0	0	0
4	5	5, 5	2	1083	17673	45387	30563	5143	151
4	6	5, 5	3	1191	17652	45527	30371	5102	157
4	7	5, 1	0	60572	39428	0	0	0	0
4	8	5, 1	1	60483	39517	0	0	0	0
4	9	5, 1	0	60232	39768	0	0	0	0
4	10	5, 1	1	60358	39642	0	0	0	0
4	11	5, 1	0	60458	39542	0	0	0	0
4	12	5, 1	1	60252	39748	0	0	0	0
4	13	5, 1	0	60418	39582	0	0	0	0
4	14	5, 1	1	60381	39619	0	0	0	0

(continued)

TABLE 6.3 **(continued)**

Species pairs		Observed	#Co	Sample null space number of co-occurrences					
				0	1	2	3	4	5
4	15	5, 1	0	60348	39652	0	0	0	0
5*	6	5, 5	0	1092	17576	45589	30386	5209	148
5	7	5, 1	0	60248	39752	0	0	0	0
5	8	5, 1	0	60442	39558	0	0	0	0
5	9	5, 1	1	60626	39374	0	0	0	0
5	10	5, 1	1	60242	39758	0	0	0	0
5	11	5, 1	1	60504	39496	0	0	0	0
5	12	5, 1	1	60597	39403	0	0	0	0
5	13	5, 1	1	60231	39769	0	0	0	0
5	14	5, 1	0	60573	39427	0	0	0	0
5	15	5, 1	0	60587	39413	0	0	0	0
6	7	5, 1	1	60620	39380	0	0	0	0
6	8	5, 1	1	60357	39643	0	0	0	0
6	9	5, 1	0	60432	39568	0	0	0	0
6	10	5, 1	0	60607	39393	0	0	0	0
6	11	5, 1	0	60567	39433	0	0	0	0
6	12	5, 1	0	60571	39429	0	0	0	0
6	13	5, 1	0	60723	39277	0	0	0	0
6	14	5, 1	1	60422	39578	0	0	0	0
6	15	5, 1	1	60496	39504	0	0	0	0
7	8	1, 1	0	93138	6862	0	0	0	0
7	9	1, 1	0	93196	6804	0	0	0	0
7	10	1, 1	0	93214	6786	0	0	0	0
7	11	1, 1	0	93116	6884	0	0	0	0
7	12	1, 1	0	93160	6840	0	0	0	0
7	13	1, 1	0	93086	6914	0	0	0	0
7	14	1, 1	0	93297	6703	0	0	0	0
7	15	1, 1	0	93222	6778	0	0	0	0
8	9	1, 1	0	93238	6762	0	0	0	0
8	10	1, 1	0	93072	6928	0	0	0	0
8	11	1, 1	0	93129	6871	0	0	0	0
8	12	1, 1	0	93196	6804	0	0	0	0
8	13	1, 1	0	93081	6919	0	0	0	0
8	14	1, 1	0	93255	6745	0	0	0	0
8	15	1, 1	0	93187	6813	0	0	0	0
9	10	1, 1	0	93277	6723	0	0	0	0
9	11	1, 1	0	93170	6830	0	0	0	0
9	12	1, 1	0	93010	6990	0	0	0	0
9	13	1, 1	0	93203	6797	0	0	0	0
9	14	1, 1	0	93131	6869	0	0	0	0
9	15	1, 1	0	93057	6943	0	0	0	0
10	11	1, 1	0	93167	6833	0	0	0	0
10	12	1, 1	0	93177	6823	0	0	0	0
10	13	1, 1	0	93148	6852	0	0	0	0
10	14	1, 1	0	93222	6778	0	0	0	0
10	15	1, 1	0	93030	6970	0	0	0	0
11	12	1, 1	0	93142	6858	0	0	0	0
11	13	1, 1	0	93305	6695	0	0	0	0
11	14	1, 1	0	93215	6785	0	0	0	0
11	15	1, 1	0	93193	6807	0	0	0	0

TABLE 6.3 **(continued)**

Species pairs		Observed	#Co	Sample null space number of co-occurrences					
				0	1	2	3	4	5
12	13	1, 1	0	93197	6803	0	0	0	0
12	14	1, 1	0	93228	6772	0	0	0	0
12	15	1, 1	0	93165	6835	0	0	0	0
13	14	1, 1	0	93340	6660	0	0	0	0
13	15	1, 1	0	93127	6873	0	0	0	0
14	15		0	93132	6868	0	0	0	0

Notes. Shown are all possible species pairs (species from 1 to 15), the observed number of times each of the two occurs, the observed number of co-occurrences (#Co), and the distribution of co-occurrences in the sample null space (Sample null space number of 0-occurrences) of 100,000 unique members for 15 species on 10 islands whose observed incidence matrix is given in table 6.1. Species pairs {(1, 2), (3, 4), and (5, 6)} co-occur significantly less often than chance expectation. The observed values—zero in each case—occur less often or equal to null distribution 1,123 times, 1,125 times, and 1,092 out of 100,000 samples, respectively An asterisk (*) marks pairs that are unusual.

not differ from chance. Connor and Simberloff had chosen an ensemble metric—and we are critical of that choice.

Connor and Simberloff (1979) used a metric that was the sum of a simple metric applied to each and every possible species pair. The response by Diamond and Gilpin was both long in coming and weak in its arguments. Many subsequent publications appeared, further entrenching the two groups.

As was pointed out by us and other authors (Stone and Roberts, Wright and Biehl, and Harvey et al.), ensemble metrics act to obscure subtle patterns exhibited in this case by some species pairs. Nevertheless, other authors followed the lead of Connor and Simberloff and introduced other ensemble metrics. Most often, these metrics demonstrated that patterns found in the community under study did not differ from chance expectation.

Diamond (1975), Harvey et al. (1983), Manly (1995), and MacNally (1990, 1989) suggested that those species pairs whose co-occurrence patterns differ significantly from chance expectations must be identified so that causal mechanisms can be investigated. Such identification depends upon a null model and metric that can distinguish chance from pattern embedded within a community. The natural metric achieves this end.

PART III

Reanalysis and Extensions

CHAPTER SEVEN

Vanuatu and the Galápagos

We regret we cannot use the same Bismarck data which Diamond first used, but its publication has been delayed by various unforeseen complications (J. M. Diamond, pers. comm.;
E. Mayr, pers. comm.). In lieu of these, we have used the New Hebridean bird data (Diamond
and Marshall 1976), plus data for West Indies birds (Bond 1971), and bats (Baker and Genoways 1978) to examine the assembly rules. — Connor and Simberloff 1979

When Connor and Simberloff could not use the Bismarck avifauna
data from which Diamond (1975) derived his assembly rules, they
turned to data on avifauna of Vanuatu that Diamond and Marshall (1976)
published. The availability of those data allowed Sanderson et al. (1998)
to reanalyze the patterns and, importantly, to develop many of the ideas
and techniques that the previous three chapters presented. In the first
part of this chapter, we present the results of this reanalysis of the birds
of Vanuatu.

In the second part of this chapter, we turn to the Galápagos. Although
neither Diamond nor Simberloff and colleagues analyzed the birds of
the Galápagos, there are excellent data on the distribution of the birds
there. These data are interesting in two important ways. We can analyze a
group of species—Darwin's finches—all derived from a common ancestor. Moreover, there are six species within just one genus. It is one that
scientists have studied particularly intensively. Second, there are detailed
phylogenies of these species that allow us to ask if checkerboard distributions are unusual among species that are most closely related. The mockingbirds on the Galápagos are a different set of species. They are the most
famous checkerboard in the history of biology.

The Birds of Vanuatu

Ernst Mayr in his *Birds of the Southwest Pacific* (1945) summarized the results of ornithological fieldwork in the region up to 1937. Mayr's book included information collected during the Whitney South Sea Expedition that visited many Pacific archipelagos from 1921 to 1939. Diamond surveyed the islands of Efate and Santo on five separate occasions between 1969 and 1976. More recently, Heinrich Bregulla (1992), who spent more than two decades in Vanuatu, published the *Birds of Vanuatu*.

Connor and Simberloff (1979) used the species and islands list given in appendix 1 of Diamond and Marshall (1976, 195–98). Stone and Roberts (1992), Sanderson et al. (1998), and others also used this same data. Bregulla (1992) assembled what is today the most accurate published description of the avifauna of Vanuatu. Bregulla updated the species distribution of the Vanuatu archipelago given by Diamond and Marshall (1976) (tables 7.1 and 7.2 in this vol.), and his species-on-islands list appears in his book (1992, 72–75). Although differences between Diamond and Marshall (1976) and Bregulla (1992) are few, they still exist. For instance, the first species in both lists, the Australian grebe (*Tachybaptus novaehollandiae*), was recorded on five islands (Gaua, Santo, Malo, Ambae [Aoba], and Eafte) by Diamond and Marshall (1976). Bregulla also found it on Maewo (Bregulla 1992). This and additional differences make comparing previously obtained results more difficult.

Note that Connor and Simberloff (1979) used Marshall and Diamond (1976) but made six errors. These errors occurred either as transcription errors or during their sorting procedure to "echelon" form in which the rows and columns were sorted in descending order. The most egregious error was one of species repetition and omission: Polynesian triller (*Lalage maculosa*) appears twice at the expense of omitting long-tailed triller (*Lalage leucopyga*). Swamp harrier (*Circus approximans*), scarlet robin (*Petroica multicolor*), and Vanuatu mountain pigeon (*Ducula bakeri*) each have a single transcription error: a presence that appears where there should be an absence, and vice versa. White-eyed duck (*Aythya australis*) has an extra absence.

As a check, when these errors were introduced into Diamond and Marshall's (1976) data, the numbers of pairs given in Connor and Simberloff (1979) are obtained exactly. Importantly, however, use of the corrected data merely perturbs the results but does not alter the conclusions reached by Connor and Simberloff (1979).

We do not use Bregulla (1992) to update Diamond and Marshall's (1976) data in tables 7.1 and 7.2, so our results are more readily comparable to those of Connor and Simberloff.

There are 24 families consisting of 45 genera with 56 species of land birds found on 28 islands of Vanuatu (table 7.1). The richest genus, *Collocalia*, has just three representatives, 9 other genera have two representatives, and the remaining 35 genera have just a single species.

Using the numbers-of-pairs metric, Connor and Simberloff (1979) presented a striking comparison of the Vanuatu birds against a sample null space derived from the observed incidence matrix (fig. 6.1). As noted previously, the fit was simply too good to be true.

Null Model Analysis

The number of presences derived from Bregulla (1992)'s species-on-islands list (table 7.2) was 888. Three species occur on all 28 islands. On the basis of the generation rate of duplicate sample null space members, Sanderson et al. (1998) estimated there were in excess of 10^{43} unique members of the full null space.

The observed number of checkerboards, 61, was close to the modal number of checkerboards among one million random incidence matrices (fig. 7.1).

Eight species accounted for the observed number of 61 checkerboards. The brown goshawk occurred only on the southernmost island, which held 32 species. Thus, the brown goshawk accounted for 24 checkerboards. Three endemics, the Santa Cruz ground dove, thicket warbler, and Santo mountain starling, all occurred on the richest island, which had 50 species. Thus, these three species added 15 checkerboards. Note that all three do not co-occur with the brown goshawk, but these three checkerboards were counted in the brown goshawk total number of checkerboards. The grey teal, which occurs on two islands, contributed seven checkerboards, the white-browed crake and the Vanuatu kingfisher, which each occurred three times, contributed four and two checkerboards, respectively.

Only three of these species, the teal, the kingfisher, and the starling, have congeners present in these islands. Three more—the ground dove, the crake, and the thicket warbler—have species broadly similar to them that are in different genera. None of these potential six pairs of species forms checkerboards.

We have argued that many checkerboards are inevitable. Their presence is simply not surprising. As if that was not enough of a concern, the

TABLE 7.1 **The Vanuatu avifauna**

Common name	Family	Genus	Species
Australian grebe	Podicipedidae	*Tachybaptus*	*novaehollandiae*
Eastern reef heron	Ardeidae	*Egretta*	*scara*
Little (Mangrove) heron		*Butorides*	*striatus*
Pacific black duck	Anatidae	*Anas*	*superciliosa*
Grey teal		*A.*	*gibberifrons*
White-eyed duck (Hardhead)		*Aythya*	*australis*
Brown goshawk	Accipitridae	*Accipter*	*fasciatus*
Swamp Harrier		*Circus*	*approximans*
Peregrine falcon	Falconidae	*Falco*	*peregrinus*
Incubator bird (Scrubfowl)	Megapodiidae	*Megapodius*	*freycinet*
Buff-banded rail	Rallidae	*Gallirallus*	*philippensis*
Spotless crake		*Porzana*	*tabuensis*
White-browed crake		*Poliolimnas*	*cinereus*
Purple swamphen		*Porphyrio*	*porphrio*
Red-bellied fruit dove	Columbidae	*Ptilinopus*	*greyi*
Vanuatu fruit dove		*P.*	*tannensis*
Pacific Imperial pigeon		*Ducula*	*pacifica*
Vanuatu mountain pigeon		*D.*	*bakeri*
White-throated pigeon		*Columba*	*vitiensis*
Rufous-brown pheasant-dove		*Macropygia*	*mackinlayi*
Green-winged ground dove		*Chalcophaps*	*indica*
Santa Cruz ground dove		*Gallicolumba*	*sanctaecrucis*
Rainbow lorikeet	Psittacidae	*Trichoglossus*	*haematodus*
Green palm lorikeet		*Charmosyna*	*palmarum*
Fan-tailed cuckoo	Cuculidae	*Cacomantis*	*pyrrhophanus*
Shining bronze-cuckoo		*Chrysococcyx*	*lucidus*
Barn owl	Tytonidae	*Tyto*	*alba*
White-bellied swiftlet	Apodidae	*Collocalia*	*esculenta*
White-rumped swiftlet		*C.*	*spodiopygia*
Uniform swiftlet		*C.*	*vanikorensis*
White-collared kingfisher	Alcedinidae	*Halcyon*	*chloris*
Vanuatu kingfisher		*H.*	*farquhari*
Pacific swallow	Hirundinidae	*Hirundo*	*tahitica*
Melanesian Cuckoo-shrike	Campephagidae	*Coracina*	*caledonica*
Polynesian triller		*Lalage*	*maculosa*
Long-tailed triller		*L.*	*leucopyga*
Island thrush	Turdidae	*Turdus*	*poliocephalus*
Scarlet robin	Pachycephalidae	*Petroica*	*multicolor*
Golden whistler		*Pachycephala*	*pectoralis*
Southern shrikebill	Monarchidae	*Clytorhynchus*	*pachycephaloides*
Broad-billed flycatcher		*Myiagra*	*caledonica*
Vanuatu flycatcher		*Neolalage*	*banksiana*
Grey fantail	Rhipiduridae	*Rhipidura*	*fuliginosa*
Spotted fantail		*R.*	*spilodera*
Fantail warbler		*Gerygone*	*flavolateralis*
Thicket warbler		*Cichlornis*	*whitneyi*
Vanuatu mountain honeyeater	Meliphagidae	*Phylidonyris*	*notabilis*
Silver-eared honeyeater		*Lichmera*	*incana*
Cardinal honeyeater		*Myzomela*	*cardinalis*
Vanuatu white-eye	Zosteropidae	*Zosterops*	*flavifrons*
Grey-backed white-eye		*Z.*	*lateralis*
Blue-faced parrotfinch	Estrildidae	*Erythrura*	*trichroa*

TABLE 7.1 **(continued)**

Common name	Family	Genus	Species
Royal parrotfinch		E.	cyaneovirens
Santo mountain starling	Sturnidae	Aplonis	santovestris
Rusty-winged starling		A.	zelandicus
White-breasted woodswallow	Artamidae	Artamus	leucorhynchus

Note. The Vanuatu avifauna is represented by 24 families consisting of 45 genera, with 56 species of land birds found on 28 islands.

number of checkerboards does not address the issue of whether the much larger number of pairs that co-occur on only one island is interesting—or on two islands, or three, etc. These are not perfect checkerboards, but their rarity might nonetheless be ecologically interesting. Simply, it is not the number of islands on which species co-occur that matters in itself. Rather, it is whether the observed number of occurrences is in any way unusual. For one species pair, a perfect checkerboard might not be unusual. Co-occurring on, say, five islands for another pair might be, as we shall soon see.

What species, if any, do form unusual patterns of co-occurrence? As we explained in previous chapters, improvements in computing power, plus a better understanding of the issues that bedevil earlier approaches, suggest that we look at species pairs individually.

Unusual Species Pairs

Excluding the three species that occur on all 28 islands, there were (53) (52)/2 = 1,378 possible species pairs. Of these pairs, 51 occurred in fewer than 5% and 22 in more than 95% of the one million null matrices (table. 7.3).

There are twelve perfect checkerboards in this list, which means some checkerboards are unusual, given the null model we use. But there are 61 perfect checkerboards in the observed data, meaning that 61 − 12 = 49 is *not* unusual. And, of course, 73 − 12 = 61 patterns of species co-occurrence are unusual, even though they are not checkerboards. So let us look at some examples.

The Vanuatu mountain pigeon (*Ducula bakeri*) and the Polynesian triller (*Lalage maculosa*) occurred nine and ten times, respectively, and co-occurred only once on 28 islands. In the sample null space of one million unique members, we found this number of co-occurrences or fewer

TABLE 7.2 **Presence of species on different islands**

Common name	Island name																													Total
	Am	An	Aw	Ef	Em	Ep	Er	F	G	L	M	Mau	Mk	ML	Mlo	Mw	Ng	O	P	Pa	S	T	Tg	Tk	Tor	U	VL	VI	
Australian grebe	0	0	0	1	0	0	0	0	1	0	0	0	0	0	1	0	0	1	0	0	0	0	0	0	0	0	0	0	5
Little (Mangrove) heron	0	0	1	0	0	0	1	0	0	0	0	0	0	0	0	0	0	0	0	0	1	0	0	0	0	0	1	0	6
Eastern reef heron	1	1	1	1	1	1	1	0	1	0	1	1	1	1	1	1	1	1	1	1	1	1	1	1	1	1	1	1	26
Pacific black duck	0	1	0	0	1	0	1	0	1	0	1	0	0	0	0	0	0	0	0	0	1	0	0	0	0	0	1	0	8
Grey teal	0	0	0	1	0	0	0	0	0	0	0	0	0	0	0	0	0	0	0	0	0	0	0	0	0	0	0	0	2
White-eyed duck	0	0	0	1	0	0	1	0	1	0	0	0	0	0	1	0	0	0	0	0	0	0	1	0	0	0	0	0	7
Brown goshawk	0	1	0	0	0	0	0	0	0	0	0	0	0	0	0	0	0	0	0	0	0	0	0	0	0	0	0	0	1
Swamp Harrier	1	1	1	1	1	1	1	1	1	1	1	1	1	0	1	1	1	1	1	0	1	1	1	0	1	1	1	0	24
Peregrine falcon	0	1	1	1	0	1	1	1	1	1	1	0	0	1	1	0	1	0	0	0	1	0	0	0	0	0	0	0	12
Incubator bird (Scrubfowl)	1	0	1	1	1	1	0	1	1	1	1	0	0	0	1	1	1	1	1	0	1	0	1	0	0	1	1	1	20
Buff-banded rail	1	1	0	1	1	1	0	0	0	0	1	0	0	0	1	1	1	1	1	0	0	1	1	0	0	0	1	0	17
Spotless crake	0	1	0	0	0	0	0	0	0	0	0	0	0	0	0	0	0	0	0	0	0	0	0	0	0	0	0	0	3
White-browed crake	0	0	0	0	0	0	0	1	0	0	0	0	0	0	0	0	0	0	0	0	0	0	0	0	0	0	0	0	3
Purple swamphen	0	1	0	1	1	1	1	0	1	1	1	1	1	0	1	1	1	1	1	1	1	1	1	1	1	0	0	0	18
Red-bellied fruit dove	1	1	1	1	1	1	1	1	1	1	1	1	1	1	1	1	1	1	1	1	1	1	1	1	1	1	1	1	28
Vanuatu fruit dove	1	1	1	1	1	1	0	1	1	1	1	0	1	1	1	1	1	1	1	1	1	1	1	1	0	0	1	1	23
Vanuatu mountain pigeon	1	0	0	0	0	0	0	0	0	0	0	0	0	0	0	1	0	1	1	0	1	0	0	0	0	1	1	0	8
Pacific Imperial pigeon	1	1	1	1	1	1	0	1	0	1	1	0	1	0	0	1	1	1	1	1	1	1	0	0	1	1	1	0	24
White-throated pigeon	1	1	1	1	1	1	1	1	1	0	1	1	1	1	1	1	1	1	1	1	1	1	1	0	1	1	1	1	25
Rufous-brown cuckoo-dove	1	0	1	1	1	1	1	1	1	1	1	1	1	1	1	1	1	1	1	1	1	1	1	1	0	1	1	1	26

Species																						
Green-winged emerald dove	1	1	1	1	1	1	1	1	1	1	1	1	1	1	1	1	1	1	1	0	27
Santa Cruz ground dove	0	0	0	0	0	0	0	0	0	0	0	0	0	0	0	0	0	0	0	0	1
Rainbow lorikeet	1	1	1	1	1	1	1	1	1	1	1	1	1	1	1	1	1	1	1	1	25
Vanuatu lorikeet	0	1	1	0	1	1	1	0	1	1	1	1	1	1	1	1	1	1	1	1	24
Shining bronze-cuckoo	1	0	0	0	1	1	1	1	1	1	1	1	1	1	1	1	0	0	0	0	12
Fan-tailed cuckoo	0	1	0	0	1	1	1	1	1	1	1	1	1	1	1	1	1	1	1	0	18
Barn owl	1	1	0	1	1	1	0	1	1	1	1	1	1	1	1	1	1	1	1	0	21
Uniform swiftlet	1	1	1	0	1	1	1	0	1	1	1	1	1	1	1	1	1	1	1	0	20
White-rumped swiftlet	1	1	0	0	1	1	1	1	1	1	0	0	0	0	0	1	0	1	0	0	11
White-bellied swiftlet	1	1	1	1	1	1	1	1	1	1	1	1	1	1	1	1	1	1	1	1	26
White-collared kingfisher	1	1	1	1	1	1	1	1	1	1	1	1	1	1	1	1	1	1	1	1	28
Vanuatu kingfisher	0	0	0	0	0	0	0	0	0	1	1	0	0	0	0	0	0	0	0	0	3
Pacific swallow	1	0	1	1	1	1	1	1	1	1	1	1	1	1	1	1	1	0	1	0	15
Polynesian triller	0	0	1	1	0	1	1	1	1	1	1	1	1	1	1	1	0	0	1	0	10
Long-tailed triller	1	1	1	0	1	1	1	1	1	1	1	1	1	1	1	1	1	1	1	1	25
Melanesian cuckoo-shrike	0	0	0	0	0	0	0	1	1	1	1	1	1	1	1	1	0	0	0	0	4
Island thrush	1	0	1	1	1	1	1	1	1	1	1	1	1	1	1	1	1	1	1	0	21
Thicket warbler	0	0	0	0	0	0	0	0	0	0	0	0	0	0	0	1	0	0	0	0	1
Melanesian gerygone	0	1	0	1	1	1	1	1	1	1	1	1	1	1	1	1	1	0	0	0	12
Spotted fantail	1	0	1	1	1	1	1	1	1	1	1	1	1	1	1	1	1	0	0	0	17
Grey fantail	1	0	1	1	1	1	1	1	1	1	1	1	1	1	1	1	1	1	1	1	25
Broad-billed flycatcher	1	1	1	1	1	1	1	1	1	1	1	0	0	1	1	1	1	0	1	1	26
Buff-bellied flycatcher	1	0	0	1	1	1	1	1	1	1	1	0	0	0	1	0	0	0	1	0	10

(continued)

TABLE 7.2 **(continued)**

Island name

Common name	Am	An	Aw	Ef	Em	Ep	Er	F	G	L	M	Mau	Mk	ML	Mlo	Mw	Ng	O	P	Pa	S	T	Tg	Tk	Tor	U	VL	VI	Total
Southern shrikebill	0	0	0	1	1	1	1	0	1	1	1	0	0	1	1	1	0	1	1	1	1	0	1	0	1	1	1	1	19
Scarlet robin	1	1	0	1	1	1	1	0	1	1	1	1	0	1	1	1	1	1	1	0	1	1	1	1	0	0	1	0	21
Golden whistler	1	1	0	1	1	1	1	0	1	1	1	1	0	0	1	1	1	1	1	0	1	0	1	1	0	1	1	0	21
White-breasted woodswallow	1	1	0	1	1	1	0	0	1	1	1	1	1	0	1	1	1	1	1	0	1	0	0	0	0	0	0	0	19
Rusty-winged starling	1	0	0	1	0	0	0	0	1	1	1	0	0	1	1	0	1	1	1	1	1	0	0	0	0	1	0	0	12
Santo mountain starling	0	0	0	0	0	0	0	0	0	0	0	0	0	0	0	0	0	0	0	0	1	0	0	0	0	0	0	0	1
Vanuatu mountain honeyeater	1	0	0	0	0	1	0	0	0	0	1	0	0	0	0	0	0	1	0	1	1	0	0	0	0	0	1	0	10
Silver-eared honeyeater	1	1	0	0	0	1	1	0	0	1	1	1	0	0	0	1	0	1	1	1	1	0	1	0	0	1	0	0	13
Cardinal honeyeater	1	1	1	1	1	1	1	1	1	1	1	1	1	1	1	1	1	1	1	1	1	1	1	1	1	1	1	1	28
Vanuatu white-eye	1	1	1	1	1	1	1	1	1	1	1	1	1	1	1	1	1	1	1	1	1	1	1	1	1	0	1	1	26
Grey-backed white-eye	0	0	1	1	1	1	0	0	1	1	1	1	1	1	1	1	1	1	1	1	1	1	1	1	1	1	1	0	26
Blue-faced parrotfinch	1	1	0	1	0	0	1	0	1	1	0	1	0	0	0	0	1	1	0	0	0	1	0	0	0	0	0	1	10
Royal parrotfinch	1	1	0	1	1	1	0	0	1	1	1	0	0	0	0	0	0	1	1	1	1	0	1	1	0	0	0	0	14
Totals for each island	38	32	20	43	35	39	40	20	36	28	45	31	16	22	41	34	34	40	36	27	50	34	33	18	19	26	34	17	888

Notes. 1 = present; 0 = absent. Key to islands: Am, Ambrym; An, Aneityum; Aw, Aniwa; Ef, Efate; Em, Emae; Ep, Epi; Er, Erromanga; F, Futuna; G, Gaua; L, Lopevi; M, Malekula; Mau, Mau; Mk, Makura; ML, Meralava; Mlo, Malo; Mw, Maewo; Ng, Nguna; O, Aoba; P, Pentecost; Pa, Paama; S, Santo; T, Tanna; Tg, Tongoa; Tk, Tongariki; Tor, Torres Group; U, Ureparapara; VL, Vanua Lava; VI, Valua.

Source. Marshall and Diamond (1976).

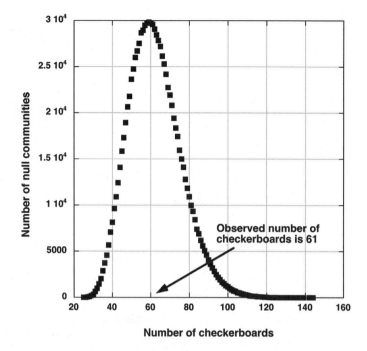

FIG. 7.1. The number of checkerboards in one million unique sample null space members ranged from 25 to 142. The observed number of checkerboards was 61.

just 6,011 times (table 7.2; fig. 7.1). Here are the numbers of null matrices with 0, 1, 2, . . . islands where the species co-occur:

0	1(observed)	2	3	4	5	6	7	8
203	5,808	53,046	197,067	340,074	279,716	106,809	16,574	703

The two species differ markedly in their feeding habits (Bregulla 1992). The pigeon eats fruit; the triller eats insects. The pigeon lives in highland forests on large islands; the triller in open woodland and gardens. Therefore there is absolutely no reason to think these species are competitively excluding each other.

Species pairs can also co-occur significantly more often than expected. For instance, the white-eyed duck (*Aythya australis*) and the Pacific black duck (*Anas superciliosa*) occurred seven and eight times, respectively, and co-occurred six times. This number of co-occurrences or more occurred in all but 989,090 of the one million null matrices. The distribution is as follows:

TABLE 7.3 **Species pairs and the number of their co-occurrences**

Genus	Species	Number	Genus	Species	Number	Co-occurrences	Sample number
Co-occur more often							
Tachybaptus	novaehollandiae	5	Aythya	australis	7	5	995503
Anas	superciliosa	8	Aythya	australis	7	6	980090
Circus	approximans	24	Columba	vitiensis	25	24	987468
Megapodius	freycinet	20	Rhipidura	spilodera	17	17	995461
Megapodius	freycinet	20	Pachycephala	pectoralis	21	19	979598
Gallirallus	philippensis	17	Tyto	alba	21	17	986509
Gallirallus	philippensis	17	Collocalia	vanikorensis	20	17	995559
Gallirallus	philippensis	17	Neolalage	banksiana	10	10	964631
Gallirallus	philippensis	17	Petroica	multicolor	21	17	986301
Pitilinopus	tannensis	23	Artamus	leucorhynchus	19	16	977837
Ducula	bakeri	8	Petroica	multicolor	21	21	992569
Ducula	pacifica	24	Phylidonyris	notabilis	10	7	982792
Collocalia	vanikorensis	20	Tyto	alba	21	21	971882
Collocalia	esculenta	26	Artamus	leucorhynchus	19	18	986457
Halcyon	farquhari	3	Lalage	leucopyga	25	25	962240
Lalage	maculosa	10	Coracina	caledonica	4	3	986933
Gerygone	flavolateralis	12	Rhipidura	spilodera	17	10	964716
Rhipidura	spilodera	17	Neolalage	banksiana	10	9	989551
Rhipidura	banksiana	10	Neolalage	banksiana	10	10	964909
Neolalage	multicolor	21	Pachycephala	pectoralis	21	17	986289
Petroica	multicolor	10	Phylidonyris	notabilis	10	8	984069
Petroica	multicolor	19	Artamus	leucorhynchus	19	18	956111
Co-occur less often							
Tachybaptus	novaehollandiae	5	Lichmera	incana	13	1	22706
Butorides	striatus	6	Megapodius	freycinet	20	3	20007
Butorides	striatus	6	Pitilinopus	tannensis	23	4	49174

Butorides	striatus	6	Macropygia	mackinlayi	26	4	4206
Butorides	striatus	6	Charmosyna	palmarum	24	4	27764
Butorides	striatus	6	Artamus	leucorhynchus	19	3	34252
Butorides	striatus	6	Eryhrura	trichroa	10	1	35911
Butorides	striatus	6	Erythrura	cyaneovirens	14	2	44758
Egretta	scara	26	Aplonis	zelandicus	12	10	32038
Egretta	scara	26	Lichmera	incana	13	11	41437
Aythya	australis	7	Lichmera	incana	13	2	25851
Accipiter	fasciatus	1	Zosterops	lateralis	26	0	17538
Falco	peregrinus	12	Megapodius	freycinet	20	7	9051
Falco	peregrinus	12	Ducula	bakeri	8	2	17106
Falco	peregrinus	12	Gerygone	flavolateralis	12	4	25946
Falco	peregrinus	12	Clytorhynchus	pachycephaloides	19	7	22519
Falco	peregrinus	12	Phylidonyris	notabilis	10	2	1979
Falco	peregrinus	12	Zosterops	lateralis	26	10	32381
Megapodius	freycinet	20	Porzana	tabuensis	3	0	646
Megapodius	freycinet	20	Poliolimnas	cinereus	3	1	28564
Megapodius	freycinet	20	Hirundo	tahitica	15	10	49784
Porzana	tabuensis	3	Lalage	maculosa	10	0	45936
Porzana	tabuensis	3	Gerygone	flavolateralis	12	0	23942
Porzana	tabuensis	3	Rhipidura	spilodera	17	0	3180
Porzana	tabuensis	3	Neolalage	banksiana	10	0	46074
Porzana	tabuensis	3	Clytorhynchus	pachycephaloides	19	1	40307
Porzana	tabuensis	3	Aplonis	zelandicus	12	0	24014
Porzana	tabuensis	3	Phylidonyris	notabilis	10	0	45572
Poliolimnas	cinereus	3	Lalage	maculosa	10	0	46059
Poliolimnas	cinereus	3	Neolalage	banksiana	10	0	45816
Poliolimnas	cinereus	3	Phylidonyris	notabilis	10	0	46013
Porphyrio	porphrio	18	Ducula	bakeri	8	4	26952
Porphyrio	porphrio	18	Chrysococcyx	lucidus	12	7	49331
Porphyrio	porphrio	18	Cacomantis	pyrrhophanus	18	10	8182
Porphyrio	porphrio	18	Lalage	leucopyga	25	15	46638

(continued)

TABLE 7.3 **(continued)**

Genus	Species	Number	Genus	Species	Number	Co-occurrences	Sample number
Porphyrio	*porphrio*	18	*Clytorhynchus*	*pachycephaloides*	19	11	18885
Porphyrio	*porphrio*	18	*Aplonis*	*zelandicus*	12	6	6307
Ducula	*bakeri*	8	*Lalage*	*maculosa*	10	1	6011
Ducula	*bakeri*	8	*Lichmera*	*incana*	13	1	475
Ducula	*pacifica*	24	*Lichmera*	*incana*	13	9	1406
Ducula	*pacifica*	24	*Erythrura*	*cyaneovirens*	14	11	43460
Chrysococcyx	*lucidus*	12	*Hirundo*	*tahitica*	15	5	13195
Chrysococcyx	*lucidus*	12	*Clytorhynchus*	*pachycephaloides*	19	7	22603
Cacomantis	*pyrrhophanus*	18	*Collocalia*	*vanikorensis*	20	12	41338
Collocalia	*esculenta*	26	*Lichmera*	*incana*	13	11	41755
Halcyon	*farquhari*	3	*Erythrura*	*trichroa*	10	0	45625
Hirundo	*tahitica*	15	*Lichmera*	*incana*	13	6	30846
Myiagra	*caledonica*	26	*Lichmera*	*incana*	13	11	41612
Clytorhynchus	*pachycephaloides*	19	*Lichmera*	*incana*	13	8	40617
Clytorhynchus	*pachycephaloides*	19	*Erythrura*	*trichroa*	10	5	5704
Phylidonyris	*notabilis*	10	*Erythrura*	*trichroa*	10	2	12245

Notes. Of 1,378 species pairs, the top 22 pairs co-occurred significantly more often than in 95% of the one million nulls, while 51 occurred fewer than in 5% of the one million nulls. "Number" indicates number of islands on which species occurred; "Co-occurrences" the number of islands on which they occurred together; and "Sample number" the number of samples, from one million, in which this number or one higher (top 22) or lower (bottom 55) occurred.

0	1	2	3	4	5	6 (observed)	7
3,012	46,912	208,783	364,578	277,053	88,752	10,567	343

There are four species of duck in Vanuatu (Bregulla 1992). Three of these in genus *Anas* are surface-feeding ducks that forage in shallow water. The Pacific black duck (*A. superciliosa*) is an example. The mallard (*Anas platyrhynchos*) is an introduced species and was not included in this analysis. The grey teal (*A. gibberifrons*) occurs only two times in Vanuatu, and its co-occurrence pattern was not unusual. The white-eyed duck (*Aythya australis*) belongs to the genus that dives to obtain food. That it shows an unusual co-occurrence with the Pacific black duck likely reflects the fact that both species use freshwater habitats. Such habitats are not present on all the islands.

One final example: there are three species of *Collocalia* swiftlets and the white-breasted woodswallow (*Artamus leucorhynchus*). All species take insects on the wing, so one might assume that the species compete for the same food and hence co-occur less often than expected by chance. In fact, one species, the white-bellied swiftlet, occurs on 26 of the 28 islands, so it must overlap extensively with the other two swifts and the woodswallow. The uniform swiftlet and the white-rumped swiftlet occur on 20 and 11 islands, respectively. They both avoid small islands so tend to occur on the same islands. (The latter occurs on one island where the former does not.)

Finally, the white-breasted woodswallow and the uniform swiftlet (*Aerodramus vanikorensis*) occur on 19 and 20 islands, respectively. The number of co-occurrences in one million unique sample null space members ranged from 11 to 19.

11	12	13	14	15	16	17	18 (observed)	19
389	8,737	64,259	212,648	341,081	265,813	93,530	13,060	483

The observed number of co-occurrences, 18 or more, occurred in more than 950,000 of the one million nulls. Again, this is because both species avoid the smallest islands (with two exceptions that are close to larger islands). According to Bregulla, the uniform swiftlet "is seen more often in open habitats than other swiftlets and is common only in the lowland and at moderate altitudes" (1992, 206). The white-breasted woodswallow "is most often seen hawking insects over forest clearings, the forest edge, park-like areas or any other type of open partly wooded country in lowland and highland regions" (273). In short, these species have different habitat preferences.

Now we return to the results as a whole. Henceforth, for simplicity, by *positively associated* or *negatively associated* species pairs, we mean pairs that co-occur on more or fewer islands, respectively, than the median values (50th percentiles) of the number of islands they co-occur generated in our 10^6 null matrices. If species were distributed randomly and independently with respect to each other, we would expect to find 5% of all the possible species pairs (that is, roughly 69 of 1,378) to be in the bottom and top 5% of the distribution. By the same reasoning, in tossing a coin six times, we expect to get six heads (or six tails) in a row less than 2% of the time for each case). Rather than 69 in each tail, there are only 22 unusual positive and 51 unusual negative pairs of species in table 7.3. These are not only the wrong numbers, but are different! Why?

Species co-occurrences are dependent. If species A and B occur on identical sets of islands, but if A and C do not co-occur on any island, then B and C also cannot co-occur. Through simulation, we can ask which species pairs occur in the lower 5% of the distribution (that is, in equal to or less than 50,000 of 10^6 runs) and, likewise, in the upper 95% of it. These pairs we call "unusual" negative or positive pairs, respectively. They are not "statistically significant at the 5% level," as one might naively be tempted to state, because we do not know the underlying statistical distribution. (That is why we must simulate it.) We can define them as "unusual" at a 5% cutoff, however.

Sample statistics depend on the distribution created by the population. Since little a priori knowledge is available regarding the distribution of the number of co-occurrences of a pair of species in the sample null space, we must employ methods that make no assumptions regarding the distribution and the parameters of the population. For instance, suppose that species A is present m times and species B n times within an archipelago of K islands. Then in each sample null community the number of co-occurrences of the species pair (A, B) lies somewhere in the interval $[\max (0, (m + n) - K), \min (m, n)]$ and varies between null communities (thus creating a distribution within the sample null space). Because the distribution of the number of co-occurrences is neither continuous nor known, we cannot employ statistics such as those used to analyze normal and other distributions. (It was routine in the earlier literature on pairwise coexistence patterns to calculate probability values on the basis of the erroneous assumption of a normal distribution.) We call statistical analyses that do not depend on knowledge of the distribution and of population parameters *nonparametric* or *distribution-free* methods.

All this means that we cannot simply read off those species that appear unusual—those in table 7.3—and make direct inferences about them. As is entirely obvious, every island set will likely produce some species pairs that will be statistically unusual.

What we can do, however, is go back to the very origins of the ideas on assembly rules and posit the following. Other things being equal, species in the same genus should be more likely to show unusually negative patterns than species not so closely related. We could even extend this statement to species in guilds, provided we defined what those guilds were a priori. So let us filter the data and ask: do we produce a richer haul of unusually negative patterns if we consider those species in the same genus?

For this island group, there are 1,387 pairs of species that might form checkerboards. Recall that we removed three species—the red-billed fruit dove, the white-collared kingfisher, and the cardinal honeyeater, which occur on all 28 islands. Obviously, these species cannot form checkerboards. This means that there are 7 pairs of congeners for the 7 species pairs in the same genus and 3 species pairs for the genus with 3 species, for a total of 10 congeneric pairs. We find, by simulation, that 51 pairs are unusually negative at the 5% cutoff. So, we would expect

$$(51/1,378) (1,378 - 10) = 50.6 \text{ unusual species pairs not in the same genus,}$$

and

$$(51/1,378) \; 10 = 0.4 \text{ unusual species in the same genus.}$$

We observe no unusual species pairs in the same genus, of course. What is telling about this calculation is that even if we were to have just one unusual species pair in the same genus, it would be two and a half times more than we would expect.

These calculations illustrate what we call "sieving" (Sanderson et al. 2009). We sieve for species in the same genus and see if the haul of unusual negative pairs is richer than expected. The conclusion for the birds on Vanuatu is no—but, importantly, this is clearly a poor place to look. Instead, we would need to find an island group where there were far more species in the same genus. That is exactly what we find in the Galápagos, and it is to that archipelago that we now turn (fig. 7.2).

The Birds of the Galápagos

As species of the same genus have usually, though by no means invariably, some similarity in habits and constitution, and always in structure, the struggle will generally be more severe between species of the same genus, when they come into competition with each other, than between species of distinct genera — (Darwin 1859)

Here, in *The Origin of Species*, Darwin spells out the basic process that leads to geographical checkerboards. He encountered one in the Galápagos—the mockingbirds—that greatly influenced his thinking. The technical work of Darwin's famous voyage in the *Beagle* appeared in the years after he returned to England in October 1836 as *The Zoology of the Voyage of H.M.S. Beagle*. The ornithologist and artist John Gould described and named the bird species that Darwin collected. Gould sorted out the taxonomic mess of both the mockingbirds and what we now call Darwin's finches. What Darwin saw and what he made of those patterns changed history. We consider the two groups in turn.

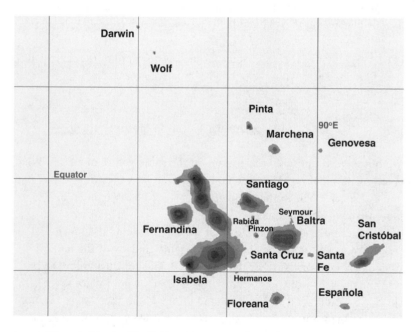

FIG. 7.2. The main islands of the Galápagos.

The Most Famous Checkerboard of All: The Galápagos Mockingbirds

In his notes, as early as 1836, the mockingbird distributions led Darwin to write that such "facts undermine the stability of species" (Sulloway 1982). The significance of what he saw grew in later editions of his writing to play a crucial role in the development of evolutionary theory. As the years passed, Darwin began to realize that the mockingbird species were likely the result of a single colonization from the mainland (where there are different species of mockingbirds) and their subsequent evolution.

The full story of which species and populations derive from which others lies beyond our scope. A recent study of the mitochondrial DNA

FIG. 7.3. The four presently recognized species of Galápagos mockingbird. Darwin knew three of them—all but that on Española. The Galápagos mockingbird occurs on Santa Cruz, Seymour, Isabela, and Ferdandina and was extirpated from Baltra (subspecies *M. p. parvulus*), Pinta (*M. p. personatus*), Genovesa (*M. p. bauri*), Santa Fe (*M. p. barringtoni*), Marchena, Santiago, and Rabida (*M. p. bindloei*), Wolf (*M. p. wenmani*), and Darwin (*M. p. hulli*). The Floreana mockingbird is extinct on Floreana but survives on two small islands nearby, one of which is Champion (<10 hectares), where this photograph was taken. The Española mockingbird occurs on that island and one small islet nearby, while the San Cristóbal mockingbird occurs on that island and one small islet nearby. Pinzon may have held a mockingbird, but if so, it is now extinct (Curry 1986). (Lower right photograph by Dušan M. Brinkhuizen and used with permission; others by S. L. Pimm)

suggests that the San Cristóbal and Española mockingbirds are similar genetically—figure 7.3 shows how similar they look—but that not all the island populations of the more widespread Galápagos mockingbird *Mimus parvulus* are closely related (Arbogast et al. 2006). These details do give some sense about how the birds have moved from island to island during their history, however.

Here is what Darwin wrote in 1839 about what we now call the Galápagos mockingbirds. We have changed the species and island names to match modern usage.

> There are five large islands in this Archipelago, and several smaller ones. I fortunately happened to observe, that the specimens which I collected in the two first islands we visited, differed from each other, and this made me pay particular attention to their collection. I found that all in Floreana belonged to *Mimus trifasciatus*; all in Isabela Island to *M. parvulus*, and all in San Cristóbal ... to *M. melanotis*. I do not rest this fact solely on my own observation, but several specimens were brought home in the *Beagle*, and they were found, according to their species, to have come from the islands as above named.

Darwin did not visit Española, which has a fourth species, *M. macdonaldi*. He continued:

> The fact, that islands in sight of each other, should thus possess peculiar species, would be scarcely credible, if it were not supported by some others of an analogous nature, which I have mentioned in my Journal of the Voyage of the Beagle. I may observe, that as some naturalists may be inclined to attribute these differences to local varieties; that if birds so different as *M. trifasciatus*, and *M. parvulus*, can be considered as varieties of one species, then the experience of all the best ornithologists must be given up, and whole genera must be blended into one species. I cannot myself doubt that *M. trifasciatus*, and *M. parvulus* are as distinct species as any that can be named in one restricted genus.

The point we wish to make is that however one divides the species, they form a perfect checkerboard: except for very small islands, every island had one mockingbird, no island had two, and several species are involved. (Sadly, the last sentence uses the past tense, since invasive species and habitat destruction has eliminated some of the island populations.)

This most famous of all checkerboards raises an immediate question about what it tells us about the process of competition in creating the observed pattern. "Why are there not two species on some islands? Surely it is because the species that got there first would exclude any subsequent invasions." The opponent replies: "The islands are so isolated from each other that different species evolved on different islands—especially on the remote ones. There need be no active process of competition at all."

These four species and their various subspecies constitute a *superspecies*—a set of allopatric (that is, geographically nonoverlapping) species clearly derived from a common ancestor. The pattern is too perfect, a little slop here and there with an island or two with two species would be more compelling, suggesting as it would that the species could move about. (Or, in the terms of the previous chapters, there is only one way in which we can arrange the incidence matrix, with one species per island.) Darwin understood the problem, carefully arguing that yes, these are clearly species (by well-established conventions), and yet they are so very close to each other geographically.

We deal with superspecies in the next chapter, for there are many sets of them in the Solomon and the Bismarck Archipelagos. There we take the conservative stance and simply lump their constituent species into one. We have not finished with the Galápagos, however.

The Search for Patterns in Darwin's Finches

Finches led Darwin to write the following: "Seeing this gradation and diversity of structure in one small, intimately related group of birds, one might really fancy that from an original paucity of birds in this archipelago, one species has been taken and modified for different ends." This quotation dates from 1845—some thirteen years before he and Wallace published their papers back to back in the *Journal of the Linnean Society* that described the all-important and thus far missing mechanism—natural selection—for how such modifications take place. This comment in the second edition of *The Origin of Species*—added to a bland description of the finches in the first edition—has gotten nearly 32,000 hits on the Internet (as of July 2013). It is surely one of the most pregnant phrases in all of science.

The finches constitute a set of species that show a remarkable array of different beaks (fig. 7.4). All this said, Darwin rather botched the finches, not realizing the need to be exact in specifying which specimens came from which islands. That, too, is another story (Sulloway 1982). We are most concerned about the pattern of their geographical distributions (table 7.4).

FIG. 7.4. Examples of six Darwin's finches. Lowland warbler finch, *Certhidea fusca*, from Española (*top left*); small tree finch, *Cactospiza parvulus*, from Santa Cruz (*top right*); small ground finch, *Geospiza fuliginosa*, from Española (*middle left*); large ground finch, *Geospiza magnirostris*, from Santa Cruz (*middle right*); vegetarian finch, *Platyspiza crassirostris*, from Santa Cruz (*bottom left*); and Common cactus finch, *Geospiza scandens*, from Santiago (*bottom right*). (Photographs by S. L. Pimm)

TABLE 7.4 **Distribution of Darwin's finches on the Galápagos**

Genus	Species	Island																
		Is	Fe	Sa	St C	Fl	Pi	Sa C	Pi	Ra	Sa F	Ma	Se	Ba	Ge	Es	Da	Wo
Geospiza	*magnirostris*	1	1	1	1	1	1	1	1	1	1	1	0	0	1	0	1	1
Geospiza	*fortis*	1	1	1	1	1	1	1	1	1	1	1	1	1	0	0	0	0
Geospiza	*fuliginosa*	1	1	1	1	1	1	1	1	1	1	1	1	1	0	1	0	0
Geospiza	*difficilis*	1	1	1	1	1	1	1	1	1	0	0	0	0	1	1	1	1
Geospiza	*scandens*	1	0	1	1	1	1	1	1	1	1	1	1	1	0	0	0	0
Geospiza	*conirostris*	0	0	0	0	0	0	0	0	0	0	0	0	0	1	1	0	0
Camarhynchus	*psittacula*	1	1	1	1	1	1	0	1	1	1	1	0	0	0	0	0	0
Camarhynchus	*pauper*	0	0	0	0	1	0	0	0	0	0	0	0	0	0	0	0	0
Camarhynchus	*parvulus*	1	1	1	1	1	1	1	0	0	0	0	0	0	0	0	0	0
Camarhynchus	*pallida*	1	1	0	1	0	1	1	0	0	0	0	0	0	0	0	0	0
Camarhynchus	*heliobates*	1	1	1	1	0	0	0	0	0	0	0	0	0	0	0	0	0
Certhidea	*olivacea*	1	1	1	1	1	1	1	1	1	1	1	1*	1*	1	1	1	1
Certhidea	*fusca*	0	0	1	0	1	0	1	1	0	1	1	0	0	0	1	1	0
Platyspiza	*crassirostris*	1	1	1	1	1	1	1	1	1	1	1	0	0	0	0	0	3
	Total species	11	10	10	10	10	9	9	9	8	8	7	4	4	4	3	3	3

Notes. Key to common names: *Geospiza magnirostris*, large ground-finch; *Geospiza fortis*, medium ground-finch; *Geospiza fuliginosa*, small ground-finch; *Geospiza difficilis*, sharp-beaked ground-finch; *Geospiza scandens*, Common cactus-finch; and *Geospiza conirostris*, large cactus-finch; *Camarhynchus psittacula*, large tree-finch; *Camarhynchus pauper*, medium tree-finch; *Camarhynchus parvulus*, small tree-finch; *Camarhynchus pallidus*, woodpecker finch; *Camarhynchus heliobates*, mangrove finch; *Platyspiza crassirostris*, vegetarian finch. We recognize two warbler finches, *Certhidea olivacea* and *Certhidea fusca*. Key to island names: Is, Isabela; Fe, Fernandina; Sa, Santiago; Sa C, Santa Cruz; Fl, Floreana; Pi, Pinzón; Sa C, San Cristóbal; Pi, Pinta; Ra, Rabida; Sa F, Santa Fe; Ma, Marcheba; Se, Seymour; Ba, Baltra; Ge, Genovesa; Es, Espanola; Da, Darwin; Wo, Wolf. An asterisk (*) indicates that the presence of *Certhidea fusca* on Seymour and Balta is suspected (on the basis of habitat), but not confirmed (Ken Petren, pers. comm.).

Darwin's finches have been the subject of many ecological, evolutionary, and morphological studies. Lack (1947)—who gave them their name—suggested that Darwin's finches exhibited *character displacement*. That is, species of finches either developed different feeding habits and hence morphological shifts in beak shape and size (and so might co-occur more often) or they simply failed to co-occur. For instance, Lack found that two species of finches with similar mandible sizes failed to co-occur on any islands.

Grant and Schluter (1984) attempted to demonstrate that the genus of ground finches, *Geospiza*, on the Galápagos Islands was structured. They showed that morphologically similar species of Darwin's finches tended not to co-occur. Connor and Simberloff (1983) rightly criticized the null community used by Grant and Schluter because the authors assumed a priori that all possible pairs of six *Geospiza* species were equally likely to occur, thus violating the row constraint. Clearly, we must avoid any a priori assumption of species co-occurrences in the null model.

We used the algorithms explained in chapters 4–6 with the observed incidence matrix (table 7.4) to create a unique sample null space of one million members, each of which satisfied the row and column constraints. For all species pairs, we recorded the number of times a particular pair of finches co-occurred in the observed incidence matrix. We compute the same metric for each species pair in each random community in the sample null space. This gives a distribution-free probability density function representing each species pair's number of co-occurrences.

If the number of times a species pair co-occurred was found in less than 5% of the sample null space, or more than 95%, then the species-pair number of co-occurrences differed from chance expectations. If not, then the number of times a species pair co-occurred did not differ from chance expectations. Of $(14 \times 13)/2 = 91$ possible species pairs, 74 pairs did not differ from chance expectations and the remaining 17 species pairs differed significantly from chance expectations (table 7.5). Five of these pairs co-occurred more often than one would expect by chance, the remaining 12 fewer than would be expected. Of the 5, one pair is congeners; of the 12, four pairs are congeners. These numbers raise several issues.

WHAT CONSTITUTES "UNUSUAL?"   First, we set the thresholds to be "unusual" at the lower 5% or upper 5% of the nulls, that is, fewer than 50,000 or more than 950,000 out of one million nulls, respectively. We would not expect there to be fewer than 5 unusual pairs in each group (5% of 91 = 4.5),

TABLE 7-5 The unusual co-occurrences of 17 pairs of finches from four genera on 17 Galápagos Islands that differed notably from that expected by chance alone

Species		Number	Species		Number	Co-occurrences	Sample number
More often than expected							
*Geospiza	fortis	13	Geospiza	fuliginosa	14	13	973836
*Geospiza	fortis	13	Geospiza	scandens	12	12	981397
Camarhynchus	psittacula	10	Platyspiza	crassirostris	11	10	987270
Camarhynchus	parvulus	10	Platyspiza	crassirostris	11	10	987397
Camarhynchus	pallida	6	Certhidea	olivacea	6	5	979766
Less often than expected							
*Geospiza	fortis	13	Geospiza	conirostris	2	0	10825
Geospiza	fortis	13	Certhidea	fusca	11	7	21679
*Geospiza	fuliginosa	14	Geospiza	difficilis	10	7	37664
*Geospiza	difficilis	10	Geospiza	scandens	12	6	44770
*Geospiza	scandens	12	Geospiza	conirostris	2	0	20432
Geospiza	conirostris	2	Platyspiza	crassirostris	11	0	36797
Camarhynchus	psittacula	10	Certhidea	fusca	11	4	441
Camarhynchus	parvulus	10	Certhidea	fusca	11	4	432
Camarhynchus	pallida	6	Certhidea	fusca	11	1	160
Camarhynchus	heliobates	2	Certhidea	fusca	11	0	36487
*Certhidea	olivacea	6	Certhidea	fusca	11	0	2
Certhidea	fusca	11	Platyspiza	crassirostris	11	5	1751

Notes. "Number" indicates the number of islands on which each species occurs, and "Co-occurrences" the number of islands on which they co-occur. "Sample number" is the number of times that that number or more extreme values were found in a sample of one million unique sample null space members. An asterisk (*) indicates that these pairs belong to the same genus.

which is what we see from the "more" group. In contrast, there are 12 pairs that have unusually few co-occurring pairs in the null space. How can this be so?

Well, the expectation that there should be fewer than 50,000 or more than 950,000 nulls is wrong. It would be correct if the co-occurrence of each pair was independent of every other pair, but this is not the case. Consider the case involving *Platyspiza crassirostris*, a species that only occurs on large islands (and by "large" recall that we mean having a large number of species, seven or more in this case). *Camarhynchus psittacula* occurs on all but one of the same set of islands, and these two species occur together more often than one would expect. *Camarhynchus parvulus* also occurs on all but one of the same set of islands. Similarly, *Platyspiza crassirostris* co-occurs with it more often than one would expect by chance.

In contrast, *Certhidea fusca* occurs on all islands with fewer than seven species (as well a few other islands). So, when one learns that *Platyspiza crassirostris* and *Certhidea fusca* co-occur on unusually few islands, it is likely that *Certhidea fusca* will show the same unusually few co-occurrences with *Camarhynchus psittacula* and *Camarhynchus parvulus* too. Table 7.5 shows that this is the case—and provides other examples.

Such relationships are of the kind: A and B are similar, A and C are different, so B and C will be different. That is, these patterns are dependent. Of course, the same constraints apply to the null space. They mean that we can readily observe more "unusual" patterns of co-occurrence than we would expect by chance, assuming independence. Importantly, this means we cannot simply read off individual patterns of unusually few co-occurrences and draw inferences about them.

Suppose we ask for just one unusual pair. The two species of *Certhidea* are very unusual. In one million nulls, we find a pattern that extreme only twice. The chance that the most unusual pair involves species in the same genus is not surprising. There are 91 pairs of species, and 26 of these are in the same genus. Two in a million is extreme, and we return to this example presently.

DATA SIEVING. Just as Darwin wrote, species in the same genus will likely be stronger competitors than less related species. So, if we sieve the pairs for those that are unusual—table 7.5—does our haul contain an unusual number of pairs in the same genus? Of the pairs that co-occur on relatively few islands, 5 involve species in the same genus, including the two species of *Certhidea* just discussed. Another 7 pairs belong to different genera. Given that 12 pairs are unusual, what is the chance that 4 of them would be congeners? Quite good, as it happens. Of the 91 possible pairs, 26 are in the same genus, 65 in different ones. By chance alone, we would expect 3.4 unusual pairs to be congeners. That is fewer than we observe, but not improbably so.

This process of sieving, however, is important. Previously we discussed sieving for Vanuatu; it will be central to our discussions of the Solomon and Bismarck Archipelagos in the next chapter, and we will see a special case of it presently.

POSITIVE ASSOCIATIONS.   If we think that congeners that rarely co-occur do so because they strongly compete and thus exclude each other, does this mean we think those that co-occur frequently are mutualists? No. They could be, of course, but there is a simpler explanation. Some islands—typically large ones—have habitats not found on smaller ones. As we have noticed above, several species occur only on large islands. Three species occur only on the eight largest islands (nine or more species) and another four species on the islands with four or more species. The critically endangered *Camarhynchus heliobates* requires tall mangrove swamps, which are only on Isabela with 11 species of birds and Fernandina with just 10 species, for example. (*Camarhynchus heliobates* is likely now extinct on the latter island.)

What Structures the Communities of Darwin's Finches?

These results beg the question: what is the structuring mechanism acting to allow the co-occurrence of some species while separating others? Using data assembled by Grant (1981, 1986) and Petren et al. (1999) after many years of study, we compare the evolutionary history, body masses, and mandible lengths, widths, depths, and feeding habits of the finches of the Galápagos Islands. Their data allow a detailed examination of the factors. We start with the genus *Geospiza*.

Consider the possibility that body mass differences explain the observed unusual co-occurrence patterns. (Body mass seems to play an important role in separating the fruit doves we discussed in chapters 2 and 3, for example.) Of the finches, *G. fortis*, *G. scandens*, *G. difficilis*, and *C. pallida*, each weighs about 20 g (Petren et al. 1999). *G. fortis* and *G. scandens* co-occur more frequently than expected by chance, while *G. fortis* and

G. difficilis, and *G. difficilis* and *G. scandens*, co-occur less frequently than expected by chance. So by itself, body mass fails to explain the finches' observed unusual co-occurrence patterns.

Darwin (1859), Lack (1947), Grant (1986), and others considered mandible dimensions as important ecological food niche indicators. Grant et al. (1985) also stated that *Geospiza* species "differ conspicuously from each other in shape, especially bill shape" and that interisland variation in food supply governed bill variation. Because average mandible width and mandible depth for each *Geospiza* across all islands were highly correlated ($r^2 = 0.998$), we restrict our analysis to examining mandible depth and mandible length.

Figure 7.5 plots the bill depths and lengths for the different island populations of the six species. The pattern is striking. The pairs of species that have unusually few co-occurrences are those that are most similar in bill size. The species pairs that co-occur unusually often are among the most morphologically different.

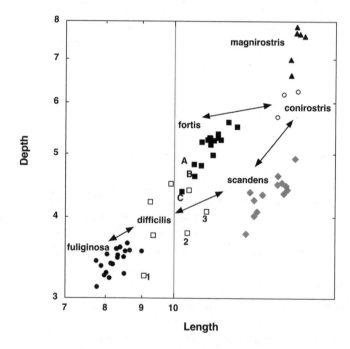

FIG. 7.5. The bill depths and lengths of different island populations of the six *Geospiza* species (from data in Grant et al. 1985). Lengths and depths are shown on a log scale. There is substantial island-to-island variation. Two-headed arrows indicate species pairs that co-occur fewer than in 5% of the null matrices. The numbers and letters refer to populations discussed in the text.

The details of the patterns are important too. The pairs that show the most obvious overlap in bill sizes between species are *G. fuliginosa* and *G. difficilis*, then *G. difficilis* and *G. fortis*.

The smallest-billed *G. difficilis* (labeled point 1 on fig. 7.5) is on an island where the generally smaller *G. fuliginosa* does not occur. Similarly, the shortest-billed *G. fortis* are on three islands (labeled points A, B, and C) where the generally smaller *G. difficilis* does not occur. Two of the three populations of *G. difficilis* with the longest bills (labeled 2 and 3) occur on islands where the generally larger *G. fortis* does not occur. These and other patterns strongly support the idea that morphologically similar species tend not to co-occur. Moreover, on islands where they do co-occur, their bill shapes are selected to be more different from their sympatric competitors than when those competitors are absent.

What about the birds' evolutionary relationships? Petren et al. (1999) analyzed phylogenetic relationships among Darwin's finches using microsatellite DNA length variation. The six *Geospiza* species are all related to each other. *G. fuliginosa* and *G. fortis* are closely related, as are *G. scandens* and *G. conirostris*. While the former pair co-occurred more often than chance expectations, the latter pair co-occurred less often than expected by chance. There is thus no definitive characteristic between evolutionary relationships and co-occurrence patterns.

The Warbler Finches, Certhidea

The two species of warbler finch show a perfect checkerboard. Every large island has one species; no island has two. Petren et al. (1999) showed that the two species of warbler finches are substantially more different from each other than are the different species of *Geospiza* and *Camarhynchus*. The warbler finches then are just like the example for the mockingbird. What makes this case so interesting is that Tonnis et al. (2005) provides a detailed phylogeny of the populations on the different islands, based on the similarities of their mitochondrial DNA. Such information provides a clue to which populations came from where.

For the mockingbirds, the genetic similarities were such that one could easily assume that greater geographic isolation translates into greater genetic differences. As Tonnis et al. (2005) write, Wallace (1880) explored this possibility in his book *Island Life*. Since then, genetic studies have demonstrated the pattern in island populations of tortoises, beetles, and lizards in the Galápagos and *Drosophila* fruit flies, spiders, and crickets in the Hawaiian Islands. Yet, for the warbler finches in the Galápagos,

144 CHAPTER 7

the reverse is true. The most distant islands are genetically similar. Figure 7.6 shows that *C. fusca* span the length and breadth of the islands. Moreover, one of those populations occurs on Hermanos, and just off the large island of Isabela, inhabited by *C. olivacea*. The two species divide the islands more or less by habitat, with *C. fusca* on low islands, *C. olivacea* on high islands, which of course have different habitats. Interestingly, there is even more genetic detail: the six populations of this species that Tonnis et al. shows are the most similar on Darwin, Wolf, Marchena, Genovesa, Española, and Hermanos. This subset itself spans the entire northwest to southeast axis of the islands.

All of these data make the evidence for active exclusion compelling. We infer that the populations can cross substantial distances—the entire archipelago, for example—so the fact that each island has only one species is not simply attributable to its isolation. As with the fruit doves in the Bismarck and Solomon Archipelagos, the ideas of Connor et al. (2013) that checkerboards are simply a consequence of geographical separation fall to the evidence presented in a good map.

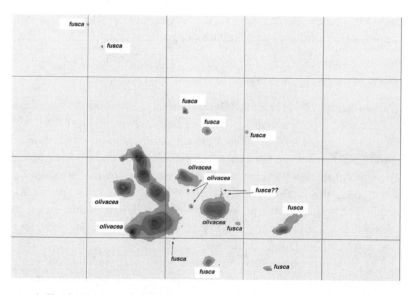

FIG. 7.6. The distribution and phylogeny of the warbler finches, *Certhidea*, on the Galápagos (from data in Tonnis et al. 2005). Geographical distance does not explain genetic similarities. *C. olivacea* occurs on generally the large, high islands. *C. fusca* occurs across the entire archipelago but on low islands. The two populations labeled *fusca*?? (on Baltra and Seymour) are likely to be this species (based on island habitat) but have not yet been assessed (K. Petren, per. comm.).

Summary

The avifauna of Vanuatu is comprised of 24 families containing 45 genera with 56 species of land birds occupying, in various combinations, 28 islands. Three species occur on all 28 islands. The total number of checkerboards did not differ from chance. A few species were responsible for most of them.

The null model analysis with one million unique sample null space communities winnowed the list of 1,387 possible pairs (= 53 × 52/2) down to a list of 73 pairs that were found to be unusual.

No unusual species pairs are in the same genus. The usual checkerboards consisted of species that were not ecologically similar. We expect both results, by chance. Indeed, even if checkerboards were two and a half times more likely among pairs in the same genus than those otherwise, we would only expect one such pair to be unusual. Simply, this is not a good system for testing whether competition structures geographical patterns.

The Galápagos, in contrast, contains two sets of species, the mockingbirds and the Darwin's finches, that have several species per genus. The mockingbirds form perhaps the most famous checkerboard in history, providing Darwin important clues to the process of evolution. Their geographical distribution, however, means that simple isolation might possibly explain which species occur where.

The group we now call Darwin's finches also influenced Darwin's thinking, even before he completed his voyage on the HMS *Beagle*. We analyzed the 17 species in 4 genera. Twelve species pairs show unusually few co-occurrences, while 5 show unusually many. Four of the 12 and 2 of the 5 are in the same genus, *Geospiza*. These patterns of co-occurrence are striking once one considers the bill sizes of the species involved. The unusually few co-occurrences are between species with the most similar bills, while the unusually many are between species with dissimilar bills. Moreover, where similar species do co-occur, they show character displacement. The bills of the smaller of the pairs are smaller, and the bills of the larger of the pairs are larger, than where the species do not overlap.

Finally, two of the finches, the warbler finches, form a perfect checkerboard. Unlike the mockingbirds, where increasing geographic distance leads to increasing genetic differences, the reverse is true for the warbler finches. The populations must have been able to cross the entire archipelago. One cannot attribute each island having only one species simply to its isolation.

The Birds of the Bismarck
and Solomon Islands

In this final chapter on island patterns, we return to the archipelagos and their birds that started this examination of geographical patterns. That there has been a technical interlude of several chapters largely reflects history. The material of this chapter follows Sanderson et al. (2009), which, in turn, relies on the data of the long-awaited monograph by Mayr and Diamond (2001). There is more than just history, of course. The preceding chapters provide a continuous argument for how to analyze geographical patterns. With those nearly thirty years of experiences in hand, what we present here is now straightforward.

Sanderson et al. (2009) assembled the observed land bird species for the Bismarck Archipelago and the Solomon Archipelago immediately to its east. The former has 150 bird species on 41 islands, with $150 \times 149/2 = 11{,}175$ possible species pairs; the latter has 141 species on 142 islands, with $141 \times 140/2 = 9{,}870$ possible species pairs. As we shall see, many of the species live in both archipelagos. Mayr and Diamond (2001) published complete records for all major islands and many small islands. To those data, Sanderson et al. added unpublished lists for 10 small Bismarck islands and 98 small Solomon Islands that Diamond and his field associates surveyed. These incidence matrices are too large to reproduce here, but they are available online with the journal *Evolutionary Ecology Research*.

The Issue of Superspecies

As did Mayr and Diamond (2001) and many other modern biogeographic studies of birds, Sanderson et al. (2009) defined the species unit of their analysis in a way other than species listed in checklists. In most cases, however, there was no distinction. For some species, we created groups called *superspecies*. So, what are they—and why did we do this? The term refers to sets of two or more populations with the following characteristics: (1) their distributions are allopatric, that is, they occupy separate islands or geographic areas; (2) the populations are believed, on morphological or molecular grounds, to be recently derived from a common ancestor; but (3) they have already achieved reproductive isolation (see Mayr and Diamond 2001, 119–26, for discussion). A good example would be the Galápagos mockingbirds of the previous chapter.

Sanderson et al. (2009) recognized 35 sets of superspecies, and figure 8.1 shows one example. The starling *Aplonis dichroa* occurs only on San Cristóbal, while three other closely related subspecies of *A. grandis* occur on island groups to the north. We simply treat these as one single unit, the superspecies—and do so exactly to address the concerns that exercised Conner et al. (2013).

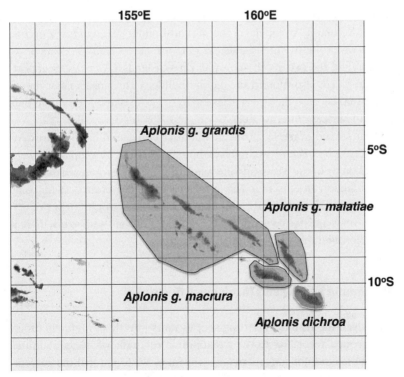

FIG. 8.1. The distribution of three subspecies of *Aplonis grandis* and of *Aplonis dichroa* in the Solomons.

Our recognition of distributions is conservative; taking the unit of analysis instead as the allospecies—"species" in the terminology used by most checklists—would inflate our results by adding 35 more perfect checkerboard distributions of (mere) allospecies. Taxonomic splitting—raising what are now considered subspecies into full species—would increase the number of checkerboards even further. Such decisions are always arbitrary, but our choice to consider, for example, all of the *Aplonis* starlings in figure 8.1 as one unit circumvents such discussions.

Thus, our analysis is unaffected by debates about whether certain closely related populations on different islands are or are not reproductively isolated, that is, whether they should be ranked as allospecies of the same superspecies or just as subspecies of the same species or allospecies. In either case, they belong to the same superspecies.

The Patterns

Figure 8.2 provides an example of the wide distributions of two species. In the Solomon Archipelago, the mound-builder *Megapodius freycinet* occurs on 110 of the 142 islands, the parrot *Eclectus roratus* occurs on 55 islands, and the two species co-occur on 55 islands. (All islands supporting the parrot also support the mound-builder.) This positive association is unusual because the two species co-occur on 55 islands or more in only 16,118 matrices of the 10^6 nulls.

Our second example shows the opposite pattern. In the Bismarck Archipelago, the nectar-feeding black honeyeater, *Myzomela pammelaena*, occurs on 23 of the 41 islands, the black sunbird, *Nectarinia sericea*, also a nectar feeder, on 14 islands. The two species never co-occur—an outcome not found in any of the 10^6 null matrices. Notice that although the first occurs on islands with very few species, and the second with many, there is still a considerable overlap in their incidences.

Taxonomic Sieving and Incidence Effects

We sieved the results by sorting species pairs into three mutually exclusive groups according to their taxonomic relatedness: nonconfamilial, confamilial but not congeners, and congeners. We found that the percentage of pairs that co-occurred significantly less than expected by chance *increased* with taxonomic relatedness.

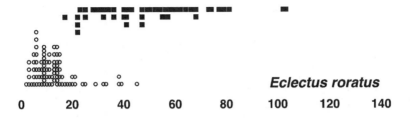

Eclectus roratus

0   20   40   60   80   100   120   140

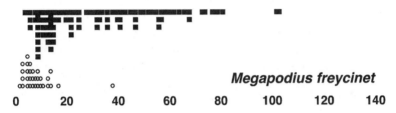

Megapodius freycinet

0   20   40   60   80   100   120   140

FIG. 8.2. The presence (black squares) or absence (open circles) of *Megapodius freycinet* and *Eclectus roratus* on islands, ranked according to their "size," that is, the total number of species of birds found on an island. From Sanderson et al. (2009).

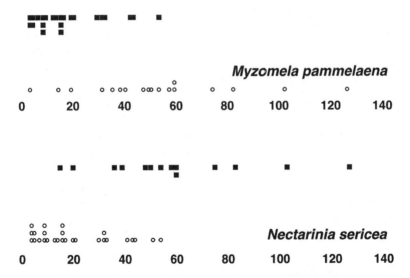

Myzomela pammelaena

0   20   40   60   80   100   120   140

Nectarinia sericea

0   20   40   60   80   100   120   140

FIG. 8.3. The presence (black squares) or absence (open circles) of the black honeyeater, *Myzomela pammelaena*, and the black sunbird, *Nectarinia sericea*, on islands, ranked according to their "size," that is, the total number of species of birds found on an island. From Sanderson et al. (2009).

For the Bismarck Archipelago, 7.7% (824 out of a possible 10,631) of nonconfamilial species pairs, 7.5% (33 out of 442) of confamilial but not congeneric species pairs, and 21.6% (22 out of 102) of congeneric species pairs co-occurred significantly less often than expected by chance (fig. 8.4). For the Solomon Archipelago the percentages increase monotonically: 6.2% (589 out of 9,434), 6.8% (23 out of 339), and 14.4% (14 out of 97) for nonconfamilial, confamilial but not congeneric, and congeneric species pairs, respectively (fig. 8.4).

The percentage of unusual pairs among congeners is highly significant statistically (p < 0.001, χ^2 test, for both archipelagos). Simply, patterns involving few geographic overlaps in distribution are most common where we would predict them to be—in closely related species. Of course, this tendency of close relatives to choose different individual islands within the same range of numbers of species on islands is what one would expect for a role of ongoing competitive exclusion, which would be strongest for the most closely related species.

We also calculated the number of observed co-occurrences for all possible species pairs and compared each to the distribution derived from the sample null space. This permits the analysis of species that are ecologically similar but taxonomically unrelated, such as pairs of honeyeaters and sunbirds (fig. 8.3). Such species pairs might co-occur less than expected by chance because of their similar, evolutionarily convergent feeding strategies. Sanderson et al. (2009) could have created a priori guilds of ecologically similar species that expand the groups beyond congeners (without the need to classify all species). They chose not to do so to fend off accusations that cherry-picking species pairs could have biased their conclusions. Nonetheless, many will find such pairs an interesting and compelling example with a plausible underlying ecological mechanism that begs further field investigation. Such deliberate exclusions make the overall results conservative.

Sanderson et al. (2009) wondered whether the high proportion of unusually negative associations among congeners arose mainly from the incidences of the pairs. For example, the fruit dove *Ptilinopus superbus* occupies 12 Bismarck islands, of which the poorest and richest support 30 and 127 species, respectively. Its congener *Ptilinopus solomonensis* occupies 22 islands, of which the poorest and richest support 4 and 83 species, respectively. The total range of islands occupied by the two species is 127 − 4 = 123; their shared range of islands is 83 − 30 = 53; and their percentage overlap in incidence is 53/123 = 43%. Yet they co-occur on

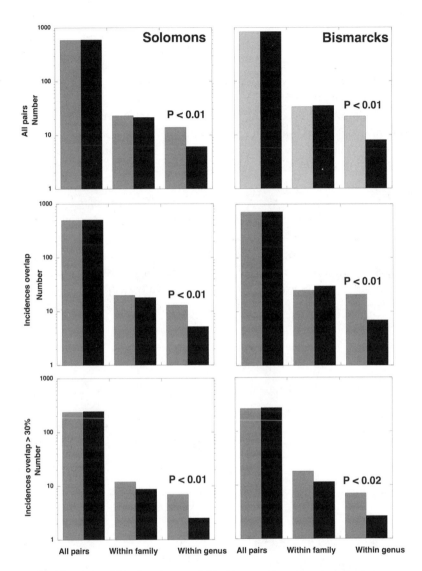

FIG. 8.4. The observed (gray) and expected (black) proportions of species pairs that are unusually negative in three mutually exclusive classes for the Solomon Archipelago (*left*) and Bismarck Archipelago (*right*). The three classes are the pairs from different families (nonconfamilial), those within the same family but not the same genus (confamilial but noncongeneric), and those within the same genus (congeneric). The three rows show all unusual pairs (*top*), the pairs that overlap in their incidences (*middle*), and those that overlap by 30% or more in their incidences (*bottom*). Note the logarithmic scale for the numbers. From Sanderson et al. (2009).

only five islands. If one species had been largely confined to islands with few species and the other to those with many that would immediately have "explained" the rareness of their co-occurrence. Instead, their incidences overlap extensively. (We will return to our insertion of "explained" in quotation marks presently.)

As figure 8.4 shows, the high proportion of unusual congeneric pairs remains high even when the species incidences overlap substantially (>30%). Simply, many congeneric pairs tend to have distributions that overlap little, *despite* having similar incidences, and not because of having differing incidences.

Consider a second example. In the Bismarck Archipelago, the fruit pigeons *Ducula rubricera* and *Ducula subflavescens* incidences overlap nearly completely (92%) and occupy islands with from 15 to 127 and from 5 to 127 species, respectively. Nevertheless, their occurrences on individual islands within this broadly overlapping range of islands are unusually exclusive, co-occurring on only six islands. In the sample null space of 10^6 communities, these species co-occur on six or fewer islands just 24,000 times (fig. 8.5).

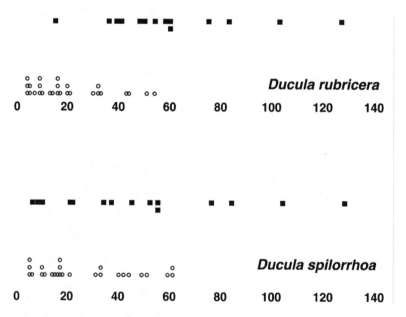

FIG. 8.5. The presence (black squares) or absence (open circles) of the pigeons *Ducula spilorrhoa* and *D. rubricera* on islands, ranked according to their "size," that is, the total number of species of birds found on an island. From Sanderson et al. (2009).

Which Genera Develop Checkerboards?

As we discussed in chapter 2, Diamond (1975) noticed a second-order assembly rule whereby the presence or absence of other species modifies
such patterns as checkerboards. In particular, larger islands than those
considered here—such as New Guinea and Borneo—often have many
coexisting congeneric species that rarely co-occur on smaller islands.
So the rule, roughly stated, might be that A and B do not co-occur, unless species P, Q, R, and S are also present, and P to S are found only on
very large islands. That specific rule is often useful in the context of larger
islands allowing species to separate by elevation—one species occurring
in the lowlands, while its similar species lives only in the mountains.

Of course, the general rule could apply in other more subtle ways.
Might the patterns of little or no geographical overlap be so sensitive to
the presence or absence of other species that the patterns of coexistence
differ between the two archipelagos? Within the islands considered here,
the answer is no—only certain genera tend to develop patterns of unusually little geographical overlap. Other genera do not. Which genera do and
which do not does not depend on the archipelago and thus on the unique
species found there.

In these islands, there are 20 genera represented by two or more congeneric species in both the Bismarck and Solomon Archipelagos. Seven
of those genera (*Ptilinopus, Ducula, Monarcha, Pachycephala, Zosterops,
Myzomela,* and *Aplonis*) generate unusual negatively associated pairs in
both archipelagos; one (*Accipiter*) does so only in the Solomon Archipelago; one (*Rhipidura*) only in the Bismarck Archipelago; and 11 (*Ardea,
Ixobrychus, Falco, Columba, Gallicolumba, Charmosyna, Micropsitta,
Aerodramus, Alcedo, Halcyon,* and *Coracina*) do so in neither. The first
set of 7 genera has 48 possible pairs in the Bismarck Archipelago and 39
in the Solomon Archipelago, of which 19 and 13, respectively, are unusual
in their few co-occurrences. The other sets of 13 genera have 42 possible
pairs in the Bismarck Archipelago and 56 in the Solomon Archipelago,
of which only one in each are unusually negatively associated. The differences between the two sets are highly significant (χ^2 test) for both the Bismarck (p < 0.0001) and Solomon (p < 0.001) Archipelagos.

Caveats

Sanderson et al. (2009) continued by considering five caveats about the data and the analyses of them, and that might conceivably alter the conclusions drawn from the results.

How Fully Known Are the Modern Avifaunas of the Islands?

If the islands were not well known, then perhaps better exploration would turn up occurrences of species that, with present knowledge, appear not to be on islands with their congeners. While undoubtedly some further distributional records will surely emerge, for birds, the Bismarck and Solomon Archipelagos are the best-explored archipelagos in the tropical Southwest Pacific Ocean. The only significant new distributional records that have appeared since the publication of Mayr and Diamond (2001) are the description of the warbler *Cettia haddeni* (LeCroy and Barker 2006), which was previously observed but not identified in the mountains of Bougainville, and reports of an unidentified *Microeca* flycatcher on New Britain and New Ireland.

To What Extent Are the Results Affected by Spatial Patterns of Species Distributions?

Put technically: does each island in a cluster of m islands, all occupied by species A and none occupied by a congeneric species B, count as one of m independent events, or must the degrees of freedom somehow be discounted? That is, if a species occupies a geographically localized subset of the Bismarck Archipelago or Solomon Archipelago while another species occupies another geographically localized subset, the spatial associations by themselves would tend to produce positive associations in some species pairs and negative associations in others. Migration might facilitate the presence of a species on different islands within such a subset of nearby islands. To count them as independent events would effectively inflate our sample sizes and hence our identifications of species pairs as unusual.

We assessed this effect by examining the contributions that the four most marked types of geographically localized distributions within the Bismarck and Solomon Archipelagos made to their 1,505 unusually negative pairs.

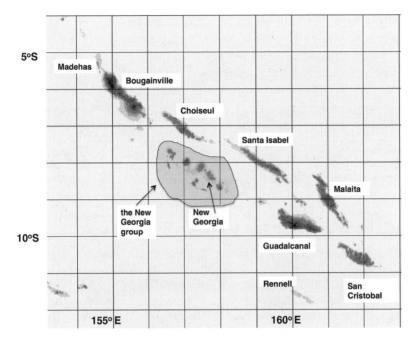

FIG. 8.6. A map showing the main Solomon Islands.

By far the most marked geographic clumping within the Solomon or Bismarck Archipelagos involves the Solomon Archipelago's New Georgia group, which contains 98 of the 142 ornithologically surveyed Solomon Islands (fig. 8.6). The group lacks 14 species otherwise geographically widespread in the Solomon Archipelago but contains 2 species confined to the New Georgia group (Mayr and Diamond 2001, 243). The two endemics are the warbler *Phylloscopus amoenus* and the white-eye *Zosterops murphyi*, both confined to the mountains of Kulambangra. The absent species also involve a white-eye *Zosterops ugiensis*, but it does not form an unusual pair with *Zosterops murphyi*. The New Georgia group also contains some endemic allospecies that do not figure separately in our analyses because they belong to more widespread superspecies. Those New Georgia group absentees and specialties are 5 of the Solomon Archipelago's 589 unusual negative pairs that do not belong to the same family. None of these pairs involves birds in the same genus. Thus they do not contribute to the result of there being more congeners than expected that show unusually little geographical overlap.

The second most marked geographic clumping involves the Northwest Bismarck Archipelago. This cluster includes 8 of the 41 ornithologically surveyed Bismarck islands and lacks 52 species otherwise geographically widespread within the Bismarck Archipelago, but they contain one species, the flycatcher *Rhipidura rufifrons*, confined in the Bismarck Archipelago to the Northwest Bismarck Archipelago (Mayr and Diamond 2001, 231–37). The Northwest Bismarck Archipelago accounts for one of the Bismarck Archipelago's 22 unusual congeneric pairs: the flycatcher *Rhipidura rufifrons* present on 2 surveyed Northwest Bismarck islands, paired with its congener *Rhipidura leucophrys* occupying 27 of the 33 Bismarck islands outside the Northwest Bismarck Archipelago. Excluding this example would not change the inference that there are more unusual congeneric pairs than expected. *R. rufifrons* is another example of a species with a wide range outside the Bismarck Archipelago, being found from the islands west of New Guinea, extensively throughout the Solomon Archipelago, and in northern and eastern Australia. It is absent from islands where *R. dahli* or *R. leucophrys* (or both) occur.

The third most marked geographic clumping involves the Solomon Archipelago's San Cristóbal group, which contains 5 of the 142 ornithologically surveyed Solomon Islands. The group lacks 9 species otherwise geographically widespread in the Solomon Archipelago but contains 5 species confined to the San Cristóbal group plus 3 other species whose Solomon distributions are concentrated in, but not restricted to, the San Cristóbal group. The group contributes none of the Solomon Archipelago's unusual negative pairs.

Finally, several unusual negative species pairs in the Bismarck Archipelago appear at first sight to have geographically complementary distributions based on their nonoverlapping incidences with one of the pair on large Bismarck islands and the other confined to outlying Bismarck islands. The *Macropygia* cuckoo doves (fig. 2.10) provide an obvious and typical example. Their segregation is not geographic: *M. mackinlayi* occurs across the entire archipelago on small, outlying islands often very close to larger islands inhabited by *M. nigrirostris*. (This latter species also occurs on some small islands.) Thus, geographical groupings make at best a minor contribution to our results.

Congeners or Guilds?

We can use the natural metric, the number of co-occurrences of a single pair of species, to decide which pairs are unusual and which are not. Our "sieving" procedure totals the number of unusual pairs according to their taxonomic relatedness. We examine the taxonomic category of congeners because we assume that congeners are more likely to compete with each other than are noncongeners. If the taxonomists have performed their job well, congeners will be more recently diverged than noncongeners. Thus, congeners are more likely to be more similar morphologically and ecologically than are noncongeners. One detailed example of how conventional taxonomy does not always get the right relationships involves the white-eyes (*Zosterops* species) just to the east of the region we consider here. Moyle et al. (2009) shows that one set of *Zosterops* species are more closely related genetically to four other genera—*Woodfordia*, *Rukia*, *Chlorocharis*, and *Stachyris*—than they are to other *Zosterops* species. In general, however, the idea of taxonomic similarity means that ecological similarity will hold. (Indeed, traditional taxonomy may be a better guide here than genetic similarities.)

This obvious fact does not deny that noncongeners may compete intensely. Some noncongeners are convergently similar and probably do compete today, a striking example being the black sunbird, *Nectarinia sericea*, and black honeyeater, *Myzomela pammelaena*, illustrated in figure 8.3. Our sieving could have chosen to focus on guilds rather than on congeners, but we did not do so because there are more uncertainties and arbitrary judgments about guild limits than about genus limits.

Simply, if we had added clear examples of ecological convergence to form guilds, such as "sunbirds plus honeyeaters," that would have enhanced our results. Given the stinging criticism, starting with Connor and Simberloff, we chose to act cautiously. In any case, had we considered guilds, whatever result we obtained would not overturn the average expectation that congeners are more likely to compete than noncongeners, and that this behavior tends to lead to checkerboard distributions.

When the Incidences Do Not Overlap

Most of the most unusual checkerboard distributions involve a "supertramp" species (a species confined to species-poor islands and absent from species-rich ones; Diamond 1975) paired with a species concentrated on species-rich islands. The *Macropygia* doves are our now-familiar example. As we noticed earlier, to suggest that nonoverlapping incidences "explain" these patterns is inadequate. Such an explanation begs the question as to why, for example, *Macropygia mackinlayi* can find its way across the Bismarck Archipelago, the Solomon Archipelago, and Vanuatu, including large islands in the Solomon Archipelago, but fails to appear on large islands in the Bismarck Archipelago. *M. mackinlayi* surely cannot be dispersal-limited, and to postulate that large islands in the Bismarck Archipelago lack some essential resource necessary for its survival smacks of special pleading. The presence of a very similar congener, *Macropygia nigrirostris*, is surely a more plausible explanation.

Simply, nonoverlapping incidences must often be the consequence of severe interspecific competition. Moreover, in the cases of superspecies—the Galápagos mockingbirds being an example—the perfect checkerboards there may also plausibly reflect the same process.

Sanderson et al. (2009) continued to note that there were previously unappreciated patterns in the taxonomic affiliations of these pairs as well as in the overlap of their incidences.

The explanation for why only certain genera tend to develop significantly nonoverlapping island distributions (negative associations) among their congeners follows from ecological differences within and between genera. Congeners co-occur on the same island or in the same archipelago through other means of niche differences. The first type is spatial. The two species may occupy different elevations—typically mountains versus lowlands—or different habitats. The second type involves nonspatial differences such that the two species have different body sizes or diets, or obtain food at different heights above the ground.

In some Northern Melanesian genera such as *Pachycephala* flycatchers and *Zosterops* white-eyes, ecological segregation is inevitably spatial. All the species in these genera are of similar body sizes, diets, and foraging techniques. In other genera, such as *Accipiter, Falco, Aerodramus,* and *Coracina,* congeners commonly overlap spatially but differ in body size, diet, or foraging technique.

From this perspective, consider the two groups discussed earlier. The 7 genera that contain pairs for which species rarely occur on the same island in both the Bismarck and the Solomon Archipelagos are ones in which ecological segregation is strictly spatial (2 genera) or predominantly so. In contrast, in the 13 genera where there is some overlap, the ecological segregation is predominantly nonspatial (9 genera) or either spatial or nonspatial (3 genera). In only one genus is it strictly spatial.

Summary

Although it has taken eight chapters to get to this point, the summary is simple. In the classic case—the bird distributions of the Bismarck and the Solomon Archipelagos—we consider species pairs that co-occur unusually. We would expect that most of these unusual co-occurrences would be ecologically uninteresting. They are.

Diamond's assembly rules expect these unusual patterns to be more common in species that belong to the same genus. They are. Moreover, they are even when the pairs involved overlap in their incidences—that is, they occur on sets of islands that have the same numbers of species on them. Even when the incidences do not overlap, the broad geographical ranges of the species often do, as the maps we have shown readily demonstrate.

Moreover, and most compelling, is the observation that seven genera produce unusual congeneric pairs *in both* the Bismarck and Solomon Archipelagos. The fact that thirteen genera do not is not troubling because species in these genera often separate ecologically by having, for example, very different body sizes.

Coda

In the Galápagos Islands and Bismarck and Solomon Archipelagos, where there are both numerous islands and species within the same genus, we find strong circumstantial evidence suggesting that ecological competition shapes which species co-occur. In Vanuatu, we did not, but then the islands have few genera with many species in them. Island systems, precisely because they are numerous and discrete, allow tests of competitive exclusion in ways that continental systems do not. If species A barely overlaps in elevation with species B, the conclusion that there is competitive exclusion is harder to make. That does not mean such exclusion is unimportant. Island systems often are informative of processes that apply elsewhere but that are harder to elucidate elsewhere. Darwin's mockingbird checkerboard is the obvious case.

There are other extensions to the work we have described thus far. These stem from the methods we have derived, and we now turn to examine those questions.

CHAPTER NINE

Species along a Gradient

The major outcome of the analyses is the dominance of nonsignificant results. — Hofer et al.
1999

In earlier chapters, we alluded to the observation that when two congeners occurred on the same island—especially the same *large* island—that
one would occur in the lowlands and the other in the mountains. The implication was that their ranges would abut (fig. 9.1b) not overlap (fig. 9.1a),
producing what we call a *parapatric* pattern. This pattern is comparable in
several ways to the island checkerboards we have discussed previously.

Gradients abound—they can be of elevation, of course, but of many
other variables too—temperature, moisture, and so on. Indeed, the ubiq-
uity of gradients suggests that looking at whether species ranges abut
should be a far more common exercise than looking for patterns on
islands. It has not been.

Another issue about gradients that cuts to the origins of plant ecology
is the very idea of a plant community. The terms to describe plant communities are so pervasive that we usually use them without thought. "Pinyon-
juniper woodlands" describes areas dominated by *Pinus edulis* and *Juni-
perus monosperma* over tens of millions of hectares in the western United
States. "Oak-hickory forest"—various oaks and hickories are involved—
stretches from New England to Georgia.

We employ these terms because they are so useful, a utility that stems
from their being shorthand for the individual communities that contain a
lot of other species, plant and animals alike. Move from one of these particular communities to the next. There is an implicit assertion that many
species ranges end with one community and start with the other. If so,
many species ranges should abut. Moreover, they should do so in more
or less the same place along the gradient (fig. 9.1d). Alternatively, sets of

species ranges could abut, but nonetheless do so in different places along the gradient (fig. 9.1c). Of course, species ranges could also be completely idiosyncratic—a pattern the figure does not show.

The mechanisms by which ranges abut could be obvious of course, reflecting a sharp boundary in physical conditions—land versus sea. Likewise, two species may end their ranges at the same place because one needs the other. Pinyon jays eat pinyon pine nuts. The patterns could also be deeper. Nearly a century ago, American plant ecologists were squabbling over whether the word *community* meant an entity in which the species fit together in some almost organic way (Clements 1916) or were "scarcely even a vegetational unit but merely a coincidence" (Gleason 1926).

The classic work of Whittaker (1967) initiated a research program aimed at understanding the structure and variation of plant communities along gradients. Whittaker suggested that certain structuring mechanisms gave rise to discernible species patterns in plant communities. Whittaker and Niering (1965, 1975), Terborgh (1971, 1985), Terborgh and Weske (1975), MacNally (1989, 1990), and subsequent researchers attempted to identify the structural mechanisms organizing communities along gradients. This group of researchers did not use the null model methodology proposed by Connor and Simberloff (1979).

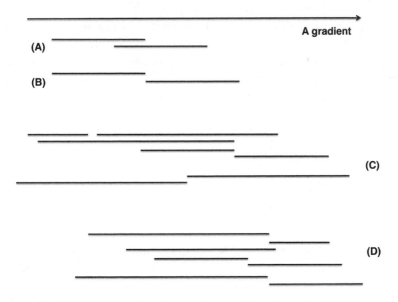

FIG. 9.1. Possibilities for the overlap of ranges along a gradient of some kind. Each line represents the interval along the gradient that a species occupies.

Certainly, the process of community assembly can progress in a way that creates alternative persistent compositions (Pimm 1991) that are difficult to invade by other species not in the mix. This book is not the place to review this field. We can ask two questions, however:

1. Do pairs of species ranges abut more often than expected by chance? (And, by friendly amendment, are these pairs the ones we expect—for example, species belonging to the same genus.)
2. Are the places on a gradient where species abut concentrated—that is, where one pair abuts, are others more likely to do so?

One cannot answer (2) without answering (1) first. We argue that (1) has not been answered well, although we provide a means to do so. We also argue that no one has yet answered question (2) with appropriate methods. There are abundant opportunities to do so.

The analysis of patterns exhibited by birds on islands has led not only to a better understanding of species co-occurrences but also to improvements in null model methods. These null model improvements have not been used to analyze species along gradients. Most of the analytic work to analyze species on gradients uses methods developed by Dale (1984, 1986, 1988), Pielou (1975, 1977, 1978), Pielou and Rutledge (1976), and Underwood (1978). These methods conserved the constraint of the number of times each species occurred but failed to conserve the additional constraint of the number of species occupying each site. Consequently, the sample null space was not representative of the problem being analyzed. As we have seen, in the absence of one less constraint, the number of unique sample null space members grows explosively, and variation within the sample null space increases because there are so many more possibilities. It is thus not surprising that interesting patterns in the observed incidence matrix were difficult to identify.

This was not the case with the herpetofauna on an elevational gradient in Cameroon (Hofer et al. 1999). Here we reanalyze Hofer at al.'s (1999) data set.

The Herptofauna of Mount Kupe, Cameroon

Previous attempts to analyze species on gradients have indeed been frustrated. The work of Hofer et al. (1999) serves as a good example. Did this frustration arise because the data needed to support a rigorous analysis were insufficient, that is, sampling omissions gave rise to erroneous results? Or was it because the method of analysis was wrong? We can test both hypotheses.

Mount Kupe is an isolated mountain, about 500 km north of the equator in the central African country of Cameroon. The mountain reaches more than 2,000 m; forests cover almost all of it. Neither of us has been there, but we readily imagine working along the slopes. It is our kind of place—rich in species, some found only there—and so a good place to test the ideas we have introduced. Let us look at what actually happened when Eric Hofer and his colleagues (Hofer et al. 1999) recorded the presence of herpetofauna on an elevation gradient from 900 to 2,000 m at 100 m intervals on Mount Kupe, Cameroon (table 9.1).

To analyze the *30x12* observed incidence matrix given (table 9.1), Hofer et al. (1999) used a traditional gradient null model, a hypergeometric model that constrained only the number of times each species occurred. They modified its use for sampling designs based on discrete points at regularly spaced intervals. Hofer et al. concluded: "The major outcome of the analysis is the dominance of nonsignificant results. The null model tests suggest that the elevational distributions of the majority of the species in the studied assemblage are limited by mechanisms other than direct competition and vegetational ecotones" (1999).

Hofer et al. (1999) found that most species co-occurrences did not differ from random expectations, though their own observations suggested some species ranges appeared to be structured. Indeed, Hofer et al. called attention to three congeneric pairs whose distributions spanned, or nearly spanned, the entire gradient and yet failed to co-occur, thus forming classic checkerboards: *Arthroleptis adolfifriderici* and *A. adelphus*, *Chamaeleo quadricornis* and *C. montium*, *Leptosiaphos* species A and *Leptosiaphos* species C. On the basis of their co-occurrence patterns, we add *Leptosiaphos* species B and C.

Hofer et al. (1999), attempting to explain the null model results, suggested that sampling errors that gave rise to absences in what should have been contiguous species ranges might have contributed to the failure of the null model to discern obvious patterns. That is, if they collected a species at 1,000 m and 1,200 m, then it must surely have occurred at 1,100 m. They did not find it there because of "sampling errors." This absence created a *hole* in the observed incidence matrix. Could sampling errors have caused catastrophic failure in the method of analysis?

TABLE 9.1 **The presence (1) and absence (0) of herptofauna species collected along an elevation gradient on Mount Kupe, Cameroon**

			Elevation											
Family	Genus	Species	900	1,000	1,100	1,200	1,300	1,400	1,500	1,600	1,700	1,800	1,900	2,000
Ar	*Arthroleptis*	*adelphus*	1	1	1	1	0	0	0	0	0	0	0	0
Ar	*Arthroleptis*	*variabilis*	1	1	1	1	1	0	0	0	0	0	0	0
Ar	*Arthroleptis*	*adolffriderici*	0	0	0	0	1	1	1	1	1	1	1	1
Ar	*Arthroleptis*	*species A*	1	1	1	1	1	1	0	0	0	0	0	0
Ar	*Astylosternus*	*montanus*	1	1	0	0	0	0	1	0	0	0	0	0
Ar	*Astylosternus*	*diadematus*	1	0	0	0	0	0	0	0	0	0	0	0
Ar	*Astylosternus*	*perreti*	1	1	1	1	1	1	0	0	0	0	0	0
Ar	*Cardioglossa*	*gracilis*	1	1	1	1	0	0	0	0	0	0	0	0
Ar	*Cardioglossa*	*venusta*	1	1	1	0	1	1	0	0	0	0	0	0
Ar	*Leptodactylon*	*ornatus*	1	1	1	1	1	0	0	0	0	0	0	0
Ar	*Trichobatrachus*	*robustus*	1	1	0	0	0	0	0	0	0	0	0	1
Ch	*Chamaeleo*	*montium*	1	1	1	1	0	0	0	0	0	0	0	0
Ch	*Chamaeleo*	*pfefferi*	0	0	1	1	1	1	1	1	1	0	1	0
Ch	*Chamaeleo*	*quadricornis*	0	0	0	0	1	0	1	1	1	1	1	1
Ch	*Rhampholeon*	*spectrum*	1	1	1	1	1	1	1	1	1	1	0	0
Sc	*Leptosiaphos*	*rohdei*	1	1	0	0	0	1	0	0	0	0	0	0
Sc	*Leptosiaphos*	*species A*	0	0	0	0	0	1	0	0	1	1	1	1
Sc	*Leptosiaphos*	*species B*	0	0	1	1	0	0	0	0	1	1	1	1
Sc	*Leptosiaphos*	*species C*	0	1	1	1	1	0	0	0	0	0	0	0
Sc	*Mabuya*	*affinis*	1	0	0	0	0	0	0	0	0	0	0	0
Ge	*Cnemaspis*	*koehleri*	0	0	0	1	1	1	1	1	1	1	0	0
Ge	*Hemidactylus*	*fasciatus*	0	1	0	0	0	0	0	0	0	0	0	0
Ge	*Hemidactylus*	*echinus*	0	0	1	1	1	0	0	0	0	0	0	0
Bo	*Calabaria*	*reinhardtii*	0	1	1	0	0	1	0	0	0	0	0	0
Co	*Bothrolycus*	*ater*	0	0	0	0	1	1	0	0	0	0	0	0
Co	*Buhoma*	*depressiceps*	1	0	1	1	0	0	1	1	1	1	1	0
Co	*Chamaelycus*	*fasciatus*	0	0	0	0	0	0	1	1	1	1	1	0
Co	*Dipsadoboa*	*species*	0	0	0	0	0	1	1	0	0	0	0	0
Co	*Mehelya*	*guiruli*	1	1	0	0	0	0	0	0	0	0	0	0
Vi	*Bitis*	*gabonica*	0	1	1	0	0	0	0	0	0	0	0	0

Notes. Forty species are included in this collection. They were collected at 900 to 2,000 m, at 100 m intervals. Abbreviations for families are as follows: Ar, Arthroleptidae; Bo, Boidae; Ch, Chamaeleonidae; Co, Colubridae; Ge, Geckonidae; Sc, Scincidae; Vi, Viperidae. Missing from this table are Arthroleptidae *Arhroleptis* sp. C, which occurred at every elevation, and nine species that were not observed during the survey hours. Species that start or end their ranges at the 1,300 m and 1,400 m elevations are underlined here and discussed in the text.

Source. From data in Hofer et al. (1999).

We can test whether or not "omissions" at some sites causes catastrophic failure of the null model. We do so below. Importantly, we ask: might an alternative null model methodology—such as the one we have applied to island communities—be used to reveal patterns in species co-occurrences along gradients that differ from random expectations? Should we add the additional constraint that species ranges be contiguous to the null model? Are there species pairs along elevation gradients that co-occur more or less often than expected by chance? The data Hofer et al. (1999) collected provide an opportunity to investigate these questions.

Hofer et al. collected data on 12 transects, each separated by 100 m between 900 and 2,000 m. Because of complex topography, transect lengths varied from 140 to 790 m. Where possible, they sampled 20 m wide strips along transects in riparian zones and 10 m wide strips along stream sides. Hofer et al. (1999, 978–79) provides more details.

Of the 64 species observed, 15 species were located outside sampling sessions and so were not assigned to specific elevations. Note that for the purpose of a null model analysis, all data collected are good data. Could it be, for instance, that some of the holes in the elevation gradient could have been filled? We do not know.

Some 40 species had elevation assignments from 900 m to 2,000 m at 100 m intervals, and we used 30 for the analysis (table 9.1). To better understand the impact of possible sampling errors, one can analyze two cases: (1) we maintained an embedded absence in a range as an absence, and (2) we assumed an embedded absence was a sampling omission and so we replaced it with a presence.

We omitted ten species from the analysis either because they were not found at any elevation surveyed or because they were found at every elevation and so contributed nothing to the analyses. Case 1 consisted of the observed data with 125 presences in a *30x12* observed incidence matrix, and case 2 with embedded absences filled and so consisted of 134 presences in a *30x12* corrected incidence matrix. Note that the column constraint implies that elevations that had the greatest number of species in the observed community retained that property in each sample null community and likewise for species occurrences. We did not use abundance data, but they proved useful in explaining the results. For each case, we generated one million unique random null communities.

In both cases, there were (30 × 29)/2 = 435 possible species pairs, of which 20 were congeneric pairs, 64 confamilial (but not congeneric) species pairs, and the remaining 351 pairs were unrelated species pairs. Once again, we can use taxonomic sieving to examine the results more closely.

For case 1 with "sampling errors" unfilled, 30% (6 of 20) congeneric pairs and 8% (24 of 415) pairs that were not in the same genus species pairs were "unusual" in that they occurred so few times. *Unusual* is in the sense that we have used the term in previous chapters. That is, the extent of overlap in the distributions occurred in no more than 5% of the one million null communities. The six congeneric pairs included both species pairs identified by Hofer et al. (1999).

Using a χ^2 test, we can consider the six unusual congeners versus the 14 that were not unusual for the pairs involving congeners (a total of 20) compared to the 34 unusual with the 381 that were not unusual (total of 415) for species that were not in the same genus. The difference is highly significant (p < 0.01), so confirming the obvious. Many more of the unusual results involve species in the same genus.

For case 2 with "sampling errors" filled, the number of unusual congeneric pairs remained the same, but some 47 noncongeneric pairs were unusual in how few times they co-occur. The excess of unusual pairs in the same genus is still significant (p < 0.03).

Hofer et al. (1999) suggested the co-occurrence pattern of only two pairs as being in any way unusual. Our results confirm Hofer at al. regarding these pairs, shown with asterisks in table 9.2. Our results also include four other unusual congeneric species pairs identified here. The high correlation of results between the two cases clearly shows that "sampling errors" do not account for the nonsignificance of Hofer et al.'s (1999) results.

These satisfying results quantify and therefore reinforce some of the suspicions of Hofer et al. (1999) that direct competition might well determine several congeneric species distributions. Hofer et al. (1999, 984) re-

TABLE 9.2 **Congeneric species pairs that occurred in fewer than 5% of the sample null space of 10^6 unique members**

Species 1	Specie	Numbers	%
Arthroleptis adelphus	*Arthroleptis adolfifriderici*	4, 8, 0	0.038
*Arthroleptis variabilis	*Arthroleptis adolfifriderici*	5, 8, 1	0.24
*Arthroleptis adolfifriderici	*Arthroleptis species A*	8, 6, 2	0.85
Chamaeleo montium	*Chamaeleo quadricornis*	4, 8, 0	0.04
Leptosiaphos species A	*Leptosiaphos* species C	7, 4, 0	0.23
Leptosiaphos species B	*Leptosiaphos* species C	6, 4, 0	0.91

Notes. The numbers in col. 3 indicate the observed number of occurrences of each species, respectively, and the observed number of co-occurrences. The percentages in col. 4 indicate the percentage of times the observed number of co-occurrences were found in the sample null space. The asterisks (*) indicate the only congeneric pairs identified by Hofer et al. (1999) as being unusual.

marked that "the few gradient distributions probably affected by direct interspecific competition, i.e., between congeners are found among the terrestrial (*Leptosiaphos*) and arboreal lizards (*Chamaeleo*) and anurans with direct development (*Arthroleptis*). Among the latter, *Arthroleptis adolfifriderici* and A. *variabilis* are most similar in morphology."

These were precisely the same species pairs whose distributions the null methodology identified as differing significantly from chance expectations. For instance, two pairs of terrestrial lizards in the genus *Leptosiaphos* co-occurred less often than expected by chance: *Leptosiaphos* species A and *Leptosiaphos* species C, and *Leptosiaphos* species B and *Leptosiaphos* species C. Hofer et al. suspected these might exhibit interspecific competition, but their null model failed to confirm their suspicions.

Some other congeneric pairs co-occurred more often than in 95% of the null models. Hofer et al. did not anticipate these.

Why Do the Results Differ from Previous Results?

Hofer et al.'s results differ from those presented here because their null space is entirely different from ours. Hofer et al. used the method of Pielou and Routledge (1976) and Underwood (1978), which conserved one of three constraints—the number of times each species occurred, but not site richness and the contiguous range of each species (Dale 1984, 1986, 1988; Pielou 1975, 1977, 1978). This method allows explosive growth of the number of possible unique members in the sample null space and effectively diluted any chance for spotting unusual patterns of co-occurrence.

Hofer et al. also used an ensemble metric to analyze the entire community. We compared all possible species pairs and found that the majority of them did not differ from chance expectations. Those co-occurrence patterns that were unexpected contributed only a *marginal difference* to the total and so the sum obscured them—just as Harvey et al. (1983, 197) suggested. As we discussed previously, the use of an ensemble metric to identify "unusual communities" actually obscures patterns.

The Second Question: Do Species Form Distinct Communities?

As figure 9.1 suggests, pairs of species ranges could abut, but do so idiosyncratically. In the context of elevation, we might see species A replace species B above 500 meters above sea level and species C replace species D above 1,000 meters above sea level. The alternative would be for pairs of species to turn over at broadly similar elevations—so that, for example, one might have one suite of species below 800 meters and another above 800 meters. How might we test for such a pattern—one that suggests that there are distinct communities?

The question "How do we show which species are found together?" is more than familiar. Go to some large ecology meeting. Probably three-quarters of the papers presented use principal components analysis, multi-dimensional scaling, or some other newfangled technique into which students and professors alike feed their data. Out pops a graph showing how things cluster, complicated enough that the explanation cures insomnia. There is no test of whether the clusters have any significance or not—a likely explanation for why the presentations seem to drag on interminably.

Figure 9.1d suggests a cluster—for the very special circumstance of a gradient. There are 23 species of lizards. Nine of these either occur at 1,300 meters or above or at 1,400 meters or below. That is, there is a narrow elevation band—200 m wide—where the species composition seems to change relatively rapidly with one set of species above it and one below. (The nine species involved are underlined in table 9.1.) Is this something unexpected? If so, it suggests that the pattern exemplified by figure 9.1d might hold.

To test this, we first filled in the gaps in the distributions of each species. So, for example, *Cardioglossa venusta* was not seen at 1,200 m but at elevations both above and below. We assumed that it would also occur at 1,200 m. Using those filled-in ranges, we ran 100,000 nulls that retained the observed row and column constraints. In addition, ranges had to be continuous—a species could not have a gap in its range. We then asked what fraction of the nulls had none or more species that started or ended their ranges in this 200 m wide interval.

We found no nulls with 10 or more, and fewer than 7% with 9. Most commonly, there were 6 (22%), 7 (38%), or 8(31%) species that started or ended their ranges in this interval. Two sets of 2 species (*Arthroleptis*, *Leptosiaphos*) that show unusually little overlap—see above—represent 4 of these 9 species.

Now, 7% is not statistically significant. It could be argued that we cherry-picked both the elevations (1,300–1,400 m) and the width (200 m). Moreover, we selected this example. So we make no inference or claim any conclusion. What we can say, however, is that what we have done provides a recipe for a much wider study of species distributions on gradients—of all kind and not just elevations. Many data sets would permit just such studies.

Summary

Previous analyses of species on gradients used a hypergeometric null model. This model constrained the number of times each species occurred but failed to conserve the number of species on each site and the species' observed range. The relaxation of constraints led to explosive growth in both the size (number) and variation in the full null space. The use of an ensemble metric assures that subtle patterns remain obscure.

The null model presented here constrained (1) the number of times each species was observed to occur and (2) the number of species observed at each site. It used the natural metric to identify those unusual species pairs whose observed number of co-occurrences differed significantly from chance expectations. In a result that parallels those of previous chapters on island patterns, there are significantly more congeneric pairs whose species ranges do not overlap than one would expect by chance. Moreover, null model results were largely invariant to absences in species ranges attributed to sampling errors.

CHAPTER TEN

Applications to Food Webs

Nestedness and Reciprocal Specialization

In all the previous chapters, our focus has been on biogeographic patterns—that is, which species live where, and, to some extent, why they
live where they live. By any definition, these patterns involve large-scale
topics in ecology. But as Pimm (1991) argued in his book *Balance of
Nature?* much ecology is small scale in another obvious way—the numbers of species a study considers. Ecologists commonly study just one
species, or at most only a few of them. One way to think about the patterns involving many species is to look at food webs. A food web is a binary matrix—which species eat which other species—and the parallels to
"which species live where" are immediately obvious.

Does this mean that we ask the same questions about food webs as we
do about biogeographic patterns? The answer is a complex mix of yes and
no. There is one pattern—nestedness—where the same tests have been
applied first to biogeographic patterns and then food webs. Nestedness is
an ensemble metric, and we have written earlier why we do not find such
measures particularly insightful. Nothing that follows disturbs this conclusion. Rather, we think that showing the preponderances of nested biogeographic patterns and nested food webs does test a similar underlying concept. We strongly suspect it is not one that most of our colleagues would
suggest if we asked them. They would say "both are nested"—and leave
it at that.

We think the deeper connection between the biogeographic patterns
we discussed earlier is with another pattern we call *reciprocal specialization*. That is an awful piece of jargon, we admit. Still, it is a familiar one:
think of those textbook examples in which a flower with some bizarrely

curved corolla exploits a hummingbird with a similarly bizarre beak to pollinate it. Each is morphologically specialized for interacting with the other.

The parallel here is not in the tests that we use in this chapter, for they are very different from those we applied to biogeographic patterns. Rather, the connection is that the search is for patterns in which some underlying process—interspecific competition in one case and co-evolution in the other—suggests particular patterns.

Nestedness

The intuition of nestedness is easy to explain. Imagine a matrix of species on islands. One island may have 50 species, then another 45, and another 44. If *all* of those 45 species are also found among the 50 species, then the species composition is nested. And if the 44 are all found among the 45 and so also among the 50, then the nested pattern holds.

What happens in practice? Figure 10.1 redraws the patterns of species on the islands of Vanuatu but shows simply black squares (as presence) and open circles (as absence) to give the visual effect. The visual impression is indeed of nestedness, but the pattern is imperfect. Certainly, the islands (columns) that have most of the species tend to have those species found on islands with fewer species, and the comparable result holds for species (rows). There are "imperfections" of both kinds: holes—species one would expect to be present on islands with many species (but are not) and outliers—species that occur beyond the set of islands on which one might expect them.

Squint at the patterns of presence and absence in this figure and imagine a curve that separates the presences from the absences. For those who are "squint-challenged," we have drawn such a curve—a parabola—that classifies about 70% of the species occurrences correctly. Were the pattern perfectly nested there would be only presences above this line and none below it. The choice of a parabola for the curve is entirely arbitrary.

To Wirt Atmar, patterns like this resonated deeply. In a long e-mail to Stuart in January 2000, he wrote enthusiastically about Carl Sagan's book *Contact*. What particularly appealed to Atmar was the way in which the extraterrestrial intelligence conveyed its message to Earth's scientists in a layered manner. There was what appeared to be an obvious, if rather uninteresting message, and noise. As the plot develops, that noise turns out to

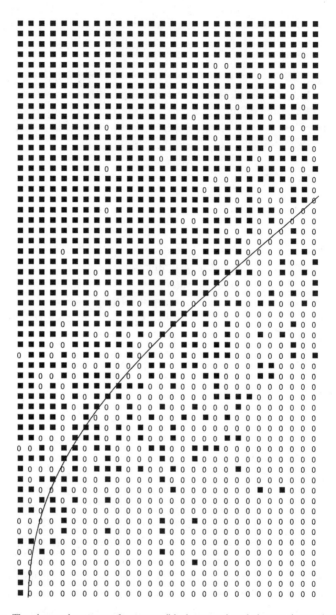

FIG. 10.1. The observed patterns of presence (black squares) and absence (open circles) of bird species on different islands in the Vanuatu group to show the approximate nested pattern. The curve suggests a line above which species occur on islands and below which they do not, were the pattern to be perfectly nested.

be full of very important information. Atmar viewed nestedness in exactly this light, something he was not able to express either eloquently or at sufficient length in the formal scientific papers he wrote with Bruce Patterson.

Atmar and Patterson wrote a series of papers on nestedness in biogeographic patterns (Atmar and Patterson 1994; Patterson and Brown 1991; Patterson 1987; Patterson and Atmar 1986; Wright et al. 1998), and Atmar provided the software to calculate a measure of nestedness. That the measure was called "temperature" was Atmar's physicist shining through; figure 10.1 looked too much like electrons in orbits to him.

Of course, when one provides computer code that anyone can turn loose on binary matrices—which are commonplace—one should not be surprised that a flood of papers follows. An immediate question would be whether the observed pattern of nestedness is what one would expect given various null hypotheses. Brualdi and Sanderson (1999) did just that. Then, of course, there was an avalanche of papers looking at technical aspects of how exactly one could measure nestedness. What we have suggested with the curve is merely suggestive of how more formal algorithms might draw the line.

That we will not explore this literature here may come as a surprise. The reason is that one expects nestedness for all kinds of good ecological reasons. Larger islands will typically have more habitats than smaller ones, even as they also include habitats that occur on smaller islands. Large and small islands have shorelines, but only large ones may have extensive areas of forest. That would be sufficient to explain nestedness. The distance of islands from the mainland could also explain patterns of nestedness: if a mainland species can reach a distant island, it can reach a near one, but the converse is not true. Either would constitute a clear signal of a well-understood process. Atmar and Patterson made that clear in their papers.

Is his correspondence, Atmar expanded by asking what ecologically important patterns might be found underneath the obvious layer of the nestedness that these processes would generate. To do that, one must move from what we have called "ensemble metrics" and look at detailed patterns of species interactions. Of course, that is just what we have done in previous chapters.

Nestedness in Food Webs

That one might look for nested patterns in food webs might seem to be an obvious next step. It is not.

Ecologists often assume a "string of beads" model for feeding relationships (Pimm 1991). In this view, each consumer tends to exploit a unique core of resources, overlapping with other consumers only in its use of less important resources. Species A overlaps with species B, B with C, C with D and so on, like beads connected on a string. Reciprocally, a unique set of consumer species threatens each resource species. Examples of niche overlaps that look like these are commonplace in ecology textbooks. Figure 10.3 in Townsend et al.'s (2003) excellent *Essentials in Ecology* is one example.

One can imagine those overlaps as being along some niche "dimension." Body size is one possible example. The largest-bodied consumers take the largest-sized resources, smaller species take smaller-sized resources, and so on.

Certainly, species may line up on some dimension such as a physical gradient—elevation, for example—as we discussed in the previous chapter. It could be the distance above or below the intertidal zone. And the dimension could also be any more abstract combination of resources (Cohen et al. 1993). This view of niche relationships underlies May's (1986) work on the limits to niche overlaps. Ecological theory and analysis that accept that two or more species *differ* in the resources they prefer (e.g., Rosenzweig 1981) encompass this "string of beads" model.

If this is the view of feeding patterns, then it is about as nonnested as one can imagine. To be nested, one consumer would eat all the available species of resources, then another a smaller subset of the same species, and so on. Such a pattern not only defies what appears to be conventional wisdom but a widely held idea that species must chose different resources if they are to coexist.

Despite the popularity of the "string of beads" model, it is likely *not* the description that most commonly applies to dietary overlaps in nature. Increasing evidence supports a very different "flower petal" model (Pimm 1991) of niche overlaps as being the most common pattern in nature (Bascompte et al. 2003; Vázquez and Aizen 2004; Bluthgen et al. 2007; Thébault and Fontaine 2008; Ings et al. 2009). Consumers exploit a shared, rather than distinct, core of resource species (the "base of the flower"), even as they exploit a few idiosyncratic resources ("the petal tips"). This model underpins ecological theory and analysis that assumes two or more species are likely to have generally the *same* preferences for the prey that they exploit (Pimm and Rosenzweig 1981).

The principal evidence for this model is that networks that show feeding relationships between consumers and their resources are indeed nested. (We have just changed from calling feeding relationships "food webs" to "networks" for a reason. Food webs cover all the feeding relationships within a community, while a network is just those species at adjacent trophic levels—consumers and their prey.) Consider the examples in figure 10.2.

Figure 10.2 shows six representations of networks. Figure 10.2a is the observed one. All six networks have the same number of species of birds (15), the species of fruits they feed on (21), and trophic interactions (51). In all but figure 10.2b, the number of fruits each bird species exploits and

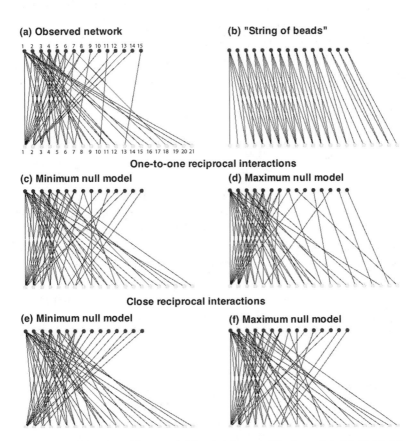

FIG. 10.2. An observed trophic network (a) and various possible alternatives (b)–(f). Models (c)–(f) have the same number of connections per consumer and per resource species, as does the observed web; model (b) does not. Given those same number of connections, models (c) and (d) have the smallest and largest number of strict one-to-one connections between consumers and resources, observed in 1,000 null models. Similarly, models (e) and (f) have the smallest and largest number of "close" pairings—those involving four or fewer species. After Joppa et al. (2009), from data in Carlo et al. (2003).

the number of birds that consume each fruit species are the same; in figure 10.2b, they are different.

Figure 10.2a, the observed network, is from the work of Carlo et al. (2003) in which the authors counted which birds ate which species of fruit in forests in Puerto Rico. The network shown is a simplification of all the data they collected, and for simplicity, we have not given the fruits and birds their scientific names. Fruit no. 1 is exploited by 11 of 15 bird species, and species no. 2 by 7 of them. Relatively specialized consumers—those that exploit only one or two species (bird species nos. 8, 10, 12, 13, and 14)—consume the resource species no. 1 or 2. Simply, most other consumers exploit these resource species as well. They are not distinctive ones, as expected under the "string of beads" model (fig. 10.2b).

In their summary, Carlo et al. drew comparable conclusions. Seven of the 68 fleshy-fruited plants were responsible for most of the birds' diets, and two species for more than half of their diets.

This particular example certainly looks nested, but is it typical? Bascompte et al. (2003) analyzed "mutualistic networks"—that is, generally mutualistic interactions between, for example, flowers and the species that pollinate them. They concluded that the majority of networks were "highly nested." Several other more recent studies have concluded the exact opposite (Ulrich and Gotelli 2007; Almeida-Neto et al. 2008). Joppa et al. (2009) revisited these conclusions, incorporating additional data and comparing the observed matrices with more tightly constrained null models of the kind we have explored in earlier chapters and applied to biogeographic problems.

Parallel arguments apply to ecological networks. Consumer species differ in how many species they exploit. The reasons why some species exploit many resources while others are more specialized are many and varied (Montoya et al. 2006). Likewise, consumer species vary in the number of species that exploit them and do so for many reasons. Any null model that does not constrain column totals creates a distribution of specialist versus generalist species that does not match nature.

Recently, several studies have used null models with row and column sums constrained to show that nestedness in several sets of ecological networks is statistically rare (Ulrich and Gotelli 2007; Almeida-Neto et al. 2008; Dormann et al. 2009; Ulrich et al. 2009), contradicting the early results of Bascompte et al. (2003).

Joppa et al. (2009) built upon Bascompte et al.'s (2003) analysis, with several modifications. First, they employed the strict null model, constraining the row and column sums, and compared their results to more loosely constrained ones. Second, they calculated nestedness using Atmar and Patterson's "temperature" (as originally done by Bascompte et al.). Then they compared those results with a more recent metric (Almeida-Neto et al. 2008) that has better statistical properties than "temperature." Finally, they extended the analysis to include antagonistic networks of parasitoids and their hosts, for comparison.

They found that row and fixed column sum constraints are sufficient to explain the patterns of nestedness in the majority of observed mutualistic and antagonistic networks. Nonetheless, they found significantly more networks than expected that were more nested than one expects by chance alone. They also found more networks that were less nested than one would expect by chance.

Whatever the underlying patterns are here, they are assembly metrics, the top, obvious layer in Atmar's analogy to signal and noise. We wish to delve deeper into that noise and look for specific patterns and the processes that shape them. The idea of coevolution provides exactly that pathway.

Ecological theories suggest that food webs might consist of groups of species forming "blocks," "compartments," or "guilds." We consider ecological networks—subsets of complete food webs—involving species at adjacent trophic levels. Reciprocal specializations occur when, say, a pollinator (or group of pollinators) specializes in a particular flower species (or group of such species) and vice versa. Such specializations tend to group species into guilds. We characterize the level of reciprocal specialization for various classes of networks. Our analyses include both antagonistic interactions—particularly parasitoids and their hosts—and mutualistic ones—such as insects and the flowers that they pollinate. We also examine whether trophic patterns might be reciprocal specializations within taxonomically related species within a network—specializations that have been obscured when the relationships were combined. We will show that reciprocal specializations are rare in all of these systems, even when tested using the most conservative null model.

Groupings of Species Interactions

From the outset, discussions of food web structure recognized the possibility of groupings of species interactions. Ecologists have variously called them guilds, compartments, or clusters. The evidence for such blocks—as May (1972) first called these groupings—is complex. Certainly, there are relatively fewer trophic interactions across habitat boundaries than within habitats (Pimm and Lawton 1980). Whether compartments or clusters exist within broadly similar habitats is less clear, with evidence both for and against (May 1973; May and McLean 2007; Montoya et al. 2006). To detect these groupings, these studies required the feeding relationships for species at three or more trophic levels (Montoya et al. 2006). Without independent information on where the groupings may be—as with those on opposite sides of habitat boundaries—one can use two trophic levels to define possible compartments and then the third to confirm them (Pimm and Lawton 1980). No one saw the means to test groupings on adjacent trophic levels before Montoya et al. (2006).

The notion that we can group feeding relationships at adjacent trophic levels is familiar (May and McLean 2007). Guilds are species that exploit similar resources, which implies that species outside the guild use distinct resources and perhaps belong to a different guild. Another perspective on these reciprocal specializations is Ehrlich and Raven's (1964) hugely influential idea of "coevolution"—the observation that taxonomically related insects might exploit taxonomically related groups of plants. In an often-cited volume, Grant and Grant (1965) referred to this as the "lock and key" model. Other examples of reciprocal patterns include hummingbirds (and other nectar-feeding birds) and the flowers they exploit (Temeles and Kress 2003), and most famously of all, the Malagasy orchid *Angraecum sesquipedale* (fig. 10.3) and the hawk moth *Xanthopan morgani*, which Darwin predicted would pollinate it (Darwin 1862). When he first saw the orchid, Darwin wrote to a colleague wondering what insect could "suck it," given its foot-long nectary. He predicted it would be a moth, although he never saw it; four decades later the pollinator was identified, and indeed Darwin was correct.

So how common are examples like this one? Joppa et al. (2009) defined strict one-to-one reciprocal interactions, as when a consumer exploits only a single resource and that resource has no other consumer—as the orchid–moth relationship is purported to be. But we should surely also count "close" reciprocal patterns whereby a small set of species—a guild—might exploit one or a small set of resources, and those resources would have no other exploiters.

There is no entirely objective measure of "close." Ehrlich and Raven's work suggested broad patterns of mutual interaction. Joppa et al. (2009) arbitrarily used four species or fewer for both consumer and resource species. Their analyses included both antagonistic interactions—particularly

FIG. 10.3. *Angraecum sesquipedale.* (Photograph by S. L. Pimm)

parasitoids and their hosts—and mutualistic ones—such as insects and the flowers that they pollinate.

While stating the question is simple, testing it is not. Joppa et al. (2009) presented two tests that differed in the expectations they assume.

If the "string of beads" model provided a commonplace description of nature, one could readily imagine how coevolution might lead to ever-tighter reciprocal specialization. For example, there might be mutual convergences in morphologies—as in the Darwin's moth–orchid case.

This "flower petal" pattern of niche overlaps might seem to preclude extensive reciprocal specialization. While eliminating or moving a few interactions in figure 10.2b could produce numerous pairs of species with reciprocal specialization, this is less easy to achieve in figure 10.2a.

Under the "flower petal" model, there need be no symmetry between resources and their consumers. For example, a pollinator may be most important to a plant but get most of its resources from a different plant (Pimm 1991; Vázquez and Aizen 2004; Bascompte et al. 2006). One can consider figure 10.2a, where three species of birds (nos. 12, 13, and 14) exploit only resource (fruit) no. 2. This fruit is likely essential to these bird species, but four other bird species also take its fruits. Collectively, those four may be more important for its seed dispersal.

In graphical terms, do observed networks more closely resemble the "string of beads" illustration (fig. 10.2b) or the nested, "flower petal" illustrations (fig. 10.2a, c–f)? We have already concluded that nature typically rejects the "string of beads" arrangement. Niche overlaps do not look like those in figure 10.2b, even in caricature.

These arguments seem to answer our question about the frequency of reciprocal trophic specialization between consumers and their resources without further analysis. No doubt the scarcity of the "string of beads" model diminishes the chances of abundant reciprocal specializations. Figure 10.2 shows that even when one constrains the numbers of connections tightly, there can be substantial variation in reciprocal specialization. Joppa et al. (2009) used exactly the same algorithms we have explored in previous chapters to test for reciprocal specialization, given the expectation that these constraints on feeding relationships will be maintained.

Figure 10.2a and c–f not only have the same number of species of birds, fruits, and interactions, but the distributions of trophic interactions per species are identical. That is, there is still a fruit species exploited by 11 bird species and another by 7. By depicting the networks in figure 10.2c–f as binary matrices, one sees that their row and column sums are the same as in figure 10.2a. Figure 10.2b is exceptional in that its row and column sums are *not* the same as the observed network.

Importantly, one can retain the broad structure of an observed network yet rearrange the trophic interactions so that there are more or fewer reciprocal mutual specializations than those observed. Figure 10.2c and d provide examples for strict, one-to-one specializations; figure 10.2e and f are for "close" specializations.

Do networks have a predominance of reciprocal specialists, given the observed distributions of consumer species using resource species, and resource species exploited by consumer species? In graphical terms, where do observed networks fall along the continuum suggested by the minimum and maximum null models shown in figure 10.2?

From the literature, Joppa et al. (2009) compiled 107 networks. They first classified the networks as those comprising mutualistic interactions (mutualistic), and those of insect hosts and their parasitoids (parasitic). In the former, the networks were mainly insects and the flowers they pollinated, or birds and the fruits they exploited.

The Overall Patterns

First, the overarching pattern is clear. Networks either show the diametrically opposite pattern to reciprocal specialization or, at the very least, are no different from what one would expect by chance in the constrained null model (table 10.1). Joppa et al. (2009) found no evidence for a preponderance of reciprocal specializations. Indeed, specialist species tend to connect to generalist species, whether one looks at consumers and their resources, or resources and their consumers. This is true using strict criteria, only looking at those interactions existing between one consumer and one resource (table 10.1, "One-to-one reciprocal interactions"), or with relaxed assumptions, looking at broad groupings of up to four consumers and four resources (table 10.1, "Close reciprocal interactions").

For either measure of reciprocal specialization, the most liberal accounting finds networks to have significantly *fewer* reciprocal specialists than one would expect (table 10.1, second to last column). However, a network cannot have fewer than zero specializations, making it impossible for original networks with zero specializations to be more specialized than expected.

Differences between Network Types

One might expect mutualistic networks to have different fractions of reciprocal specializations than antagonistic ones have. In the former, the resource wants to be eaten—at least by the right species—to disperse its pollen or fruit. In the latter, it does not. Various others have considered the different trade-offs. Table 10.1 tests the differences. For one-to-one specialization, the 15% (7 versus 39) of mutualistic networks that are more specialized than expected do not differ significantly from the 4% (1 versus 25) of parasitic networks (χ^2 test, p = 0.30). The comparable percentages for close specialization are 19% (11 versus 48) for mutualistic networks and 34% (11 versus 21) of parasitic networks, and similarly do not differ significantly (p = 0.16).

Palimpsests

Joppa et al. (2009) also examined whether trophic patterns might be "palimpsests"—that is, whether there might be reciprocal specialization within taxonomically related species within a network but that these might be obscured when these relationships are combined. They created

TABLE 10.1 **Results for all original and daughter networks included in the analysis**

One-to-one reciprocal interactions

	Total	Reciprocal	Subtotals	Equal	Less	More
Mutualistic	65	0	49	19	30	
		1	12	0	6	6
		>1	4	0	3	1
Parasitic	42	0	40	17	23	
		1	1	0	1	0
		>1	1	0	0	1
Daughter	35	0	16	4	12	
		1	8	0	3	5
		>1	11	2	5	4
Sum all networks					83*	17
Sum networks >0 specializations					18	17

Close reciprocal interactions

	Total	Reciprocal	Subtotals	Equal	Less	More
Mutualistic	65	0	6	3	3	
		1 to 10	35	1	26	8
		11 to 20	14	2	11	1
		>20	10	0	8	2
Parasitic	42	0	6	5	1	
		1 to 10	21	4	9	8
		11 to 20	9	1	6	2
		>20	6	0	5	1
Daughter	35	0	0			
		1 to 10	18	10	6	2
		11 to 20	16	13	3	0
		>20	1	1	0	0
Sum all networks					78*	24
Sum networks >0 specializations					74*	24

Notes. Col. 1 (Total) indicates the number of networks analyzed; col. 2 (Reciprocal) indicates the number of specializations present in the original network; and col. 3 (Subtotals) indicates the number of networks in each of the three categories in col. 2. Cols. 4, 5, and 6 categorize the observed networks into those that are no different from expectation (Equal), less specialized (Less), or more specialized (More). See the text for the determination of categories. Where "Reciprocal" equals zero, the column "More" is blank by default, as it is impossible for a null network to have less than zero specializations. The asterisk (*) indicates significance under a Sign test at $p = 0.05$.

Source. After Joppa et al. (2009).

35 taxonomically similar "daughter networks" from the five largest mutualistic networks. When creating these daughter networks, they retained only networks that contained greater than five insect species (of the same family) and five plant species. These daughter networks are overlapping or "overwritten" subnetworks within the original, mother networks.

Figure 10.4 provides an example. Inouye and Pyke (1988) counted the insects that came to flowers in an alpine habitat of the Snowy Mountains in New South Wales, Australia. They counted a total of 60 species flies (diptera) and smaller numbers of bees and other insects. The complete mother network shows the now-familiar pattern of most of the consumer species feeding on a very few plant species. Considering "close" interactions for this example, we find that 1.7% of the sample null networks are the same as those observed, 97% have more specialized reciprocal interactions, and only 1.3% fewer reciprocal interactions. For one-to-one

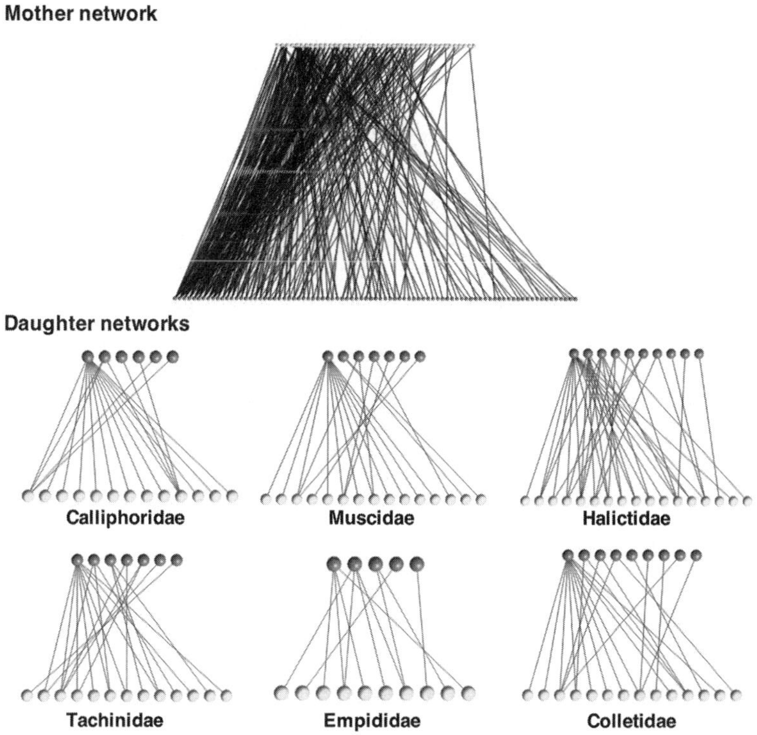

FIG. 10.4. An example of an observed mother network and its six resulting daughter networks. Of 28 insect families, only 6 met the criteria to be analyzed as a daughter network. After Joppa et al. (2009) from original data in Inouye and Pyke (1988).

interactions, the percentages are 39%, 20%, and 40%, respectively. This mother network follows the general pattern of showing fewer reciprocal specializations than expected.

The four fly families (Calliphoridae, Muscidae, Tephritidae, Empididae) and two bee families (Halictidae, Colletidae) have different feeding ecologies, of course. So Joppa et al. posited that complete networks might be composed of smaller and overwritten networks, where patterns of reciprocal specializations within a taxonomically related set of species (those in the same family, for example) might nonetheless overlap in complex ways and obscure those original patterns.

For "close" interactions, 100% of the nulls for all daughter networks are the same as those observed, so we can draw no inferences. For one-to-one interactions, only one daughter network has 100% of the nulls the same as observed (Colletidae). Four daughter networks are more reciprocally specialized than expected, while one, Muscidae, is the reverse. The Empididae network is the most reciprocally specialized subnetwork, with 58% of null models less reciprocally specialized and 42% equal to the observed network.

Table 10.1 compares the percentages of mutualistic mother networks with those of the daughter networks for several examples. For one-to-one specialization, the 31% of daughter networks (9 versus 20) that are more specialized than expected is higher, but not significantly so (χ^2 test, p = 0.18), than the mother networks (7 versus 39). For close specializations, the percentages are essentially the same (18% daughters, 2 versus 9; 19% mothers, 11 versus 48; χ^2 test, p = 0.70).

Figure 10.4 is a handpicked example, of course, but it is particularly rich in both overall species and daughter networks. Joppa et al. (2009) concluded, however, that "the contrast between the mother network, showing no evidence of mutual specializations, and the few daughter networks that do, suggests a more extensive survey of species-rich networks is required."

Summary

This chapter covers two ideas. The more specialized one asks whether trophic networks have a preponderance of reciprocal specialists. The famous example is the moth whose existence Darwin predicted (Darwin 1862). (Even here the story may not as simple as it seems, however; see Wasserthal 1998.) We found no evidence that reciprocal specialization shapes ecological networks by creating an excess of reciprocally specialized trophic interactions between consumers and resources. Joppa et al. (2009) went to particular lengths to find such reciprocal interactions, comparing networks with mutualistic interactions and those with parasitoid–host interactions. There are no compelling differences, even when correcting for the role of nestedness in network structure. Indeed, Joppa et al. found ecological networks no more nested than expected given our most conservative null model. By analogy to a palimpsest, they selected taxonomic subsets to produce daughter networks from larger networks to test whether overlapping taxonomic groupings obscured reciprocal specialization. In general, it did not.

The second idea involves nestedness—a pattern that likely applies to both biogeographic distributions and trophic networks. In the former, it would be hard to imagine that the pattern would not hold. Differences in island size, the habitats they offer species, and the distance from source areas would all conspire to give a nested pattern of species occurrences. Indeed, such a pattern may be least likely to occur when such factors are arranged to create trade-offs.

For trophic networks, conventional wisdom rejects nestedness, arguing instead that species dietary preferences should be distinct and strung along some gradient like a "string of beads." Despite textbooks' predilection for this model, almost everyone who examines feeding relationships in the field rejects it. These results suggest that common processes and universal constraints might operate across such widely different networks, and lend support to the flower petal model of network structure (Pimm 1991).

The likely explanation is surely no more complicated than variation in species abundance. Most consumer species likely exploit the most abundant resource species, while abundant consumers may be the greatest threat (or benefit, if mutualists) to resources.

In both cases, we find nestedness—an ensemble metric—simply a means of characterizing patterns with obvious underlying causes. The real interest lies in the patterns lurking beneath them, exactly as envisioned by Wirt Atmar when he first characterized nestedness.

CHAPTER ELEVEN
Coda

Why do we find the material in the previous ten chapters sufficiently important to write this monograph about them? The answer is certainly not to resolve the long and often acrimonious dispute between two highly respected groups that is now into its fifth decade. Rather, we suggest that the following points, stretching from the general to the particular, merit the attention.

MacArthur's Original Vision

As we discussed in the introduction, MacArthur's research program aspired to a bold vision whereby we might understand observed patterns in nature, or succinct "laws." That there are such generalities is not new. Wallace promulgated his famous Sarawak law (1855, 24).

Two very influential Oxford ecologists—Elton and Lack—published books on communities: *The Pattern of Animal Communities* (Elton 1966) and *Ecological Isolation in Birds* (Lack 1971). The former is largely descriptive, the latter rich in numbers, but neither had much synthesis or underlying theory. They are far from these authors' most cited works, and MacArthur's *Geographical Ecology* (1972) overshadows them.

MacArthur argued that there are simple mathematical descriptions that capture the essence of species distributions. Underpinning the patterns of species co-occurrence are the familiar competition equations, a pair of differential equations usually viewed as giving the conditions under which species can coexist. Importantly—perhaps even surprisingly—they contain conditions where the outcome of which species excludes the other depends on which gets there first. That is how one generates checkerboards, of course.

Now, there are many laws describing ecological communities (Lawton 1999), so the absence of checkerboards does not harm their existence. It does, however, diminish the vitality of a research program that seeks laws versus one that ascribes what we see as random. So, the debate this book discusses is part of a wider one about the utility of different approaches to understanding communities.

The Patterns Themselves

We find the patterns intriguing for several reasons. This is not because we are surprised by competition as a process. There is abundant evidence that similar species compete. What is interesting is the scale and extent over which competition plays out. As this debate unfolded, some asked: "what's wrong with doing an experiment?" Both sides rejected such an approach vehemently.

"Just because you can calculate this gravity force thing with your pendulum does not mean it is what is causing the planets to go around the sun!" we wished to respond. Likewise, those who demonstrated that species competed in tiny experimental plots impressed no one other than the experimenters. At issue was not whether competition is *intensive* enough to affect species abundance at small scales, but rather was it *extensive* enough to shape the large-scale patterns of nature? Our maps discuss patterns across millions of square kilometers.

Our impression of papers in community ecology was that this debate led to small-scale, experimental studies with fewer species, eschewing the issues at larger scales (Pimm 1986). If so, that would be unfortunate. Biogeographical patterns have exercised considerable historical significance. The checkerboard of the Galápagos mockingbirds and patterns of mutual exclusivity in the Amazon and insular Southeast Asia were vital clues that gave Darwin and Wallace their ideas about evolution.

The famous mockingbirds were diagrammatic. Such a striking pattern could not be ignored. The mockingbirds helped uncover the role of evolution for situations that were not so obvious. Similarly, the patterns of species co-occurrences we demonstrate have implications far beyond the islands off New Guinea. Simply, one does not need to find checkerboards everywhere. They hint at a geographically and taxonomically broader role for competition shaping species distributions in places where—again!—patterns may not be so obvious.

Human actions are now driving species to extinction at 1,000 times their natural rate (Pimm et al. 2014), so we may soon lose the clarity that gave Darwin and Wallace their ideas and give us the tests of the geographical extensive role of competition as a process. There is a further conservation message, of course. Checkerboards speak to historical contingency: whichever species gets there first wins. As we mix up the planet's flora and fauna, we give a substantial edge to "getting there first" to the species that are our commensals, those that can readily travel with us—as rats in a ship's hold or spiny seeds on our clothes, or because we think they are otherwise desirable in places where we live and they once did not.

The other patterns we discuss have other implications for conservation. Chapter 9 discusses species along gradients. Globally, climate disruption is forcing species to higher elevations, though slower than species would travel were they to stay within their recent climatic bounds (Forero-Medina et al. 2011). Does the movement of one species change the likely elevational limit of others? The degree to which species ranges abut is vital to understanding this question.

The importance of nestedness (chap. 10) for setting conservation priorities is well understood and entirely obvious. If species distributions are entirely nested, then those places with more species—typically simply larger areas—will contain all the species found in smaller ones. Similarly, large food webs will have all the components found in smaller ones. Whether either is not nested, then small areas or small webs contain unique species or patterns of interactions that we might seek to protect. Chapter 10 also discuss reciprocal specializations. One worry for those who manage species is the consequence of losing a species to the other species that depend on it in some way.

The Need for Null Hypotheses

Stuart first heard Simberloff speak at the meeting of the Ecological Society of America meeting in Amherst, Massachusetts, in the summer of 1973. Simberloff's talk was outstanding, and Stuart remembers the presentation vividly. Simberloff discussed the mangrove experiments and a variety of claims made about the assembly of island communities. Carefully crafting a series of alternative hypotheses, Simberloff demonstrated that most of the observations could be explained by chance rather than by the strong, if invisible, hand of the interactions between species.

At that time, there seemed to be a wealth of papers that uncritically accepted a wide variety of ecological patterns with no thought given to whether or not the patterns could have arose randomly. One popular genre was to assess the hypothesis that rare species are more specialized— that they have "narrower niches." Almost inevitably, the sample sizes for rare species were smaller than those for common species, and with smaller sample sizes, species inevitably appeared more specialized. Simberloff's important contribution was to unleash an army of graduate students and fellow travelers to savage anyone foolish enough to claim a result without testing against an appropriate null hypothesis.

A central theme of our book is the complexity of setting up and testing null hypotheses for complex ecological patterns. This book—and certainly its central chapters—show how very difficult it can be to set up plausible null hypotheses. Moreover, even when one does, the metric one uses to explore its behavior must be suited to the task. "Cloaking metrics" will not uncover patterns, even when interesting patterns lurk. One might wish for a simpler world, but testing patterns of species co-occurrence, nested-ness, gradients, and reciprocal specialization are topics that will not yield to simple-minded tests. That we have included all of them here is because they are all characterized by *binary matrices*, matrices of zeros and ones— clearly a very common class of problems.

Finally, there is the quite general problem of how much a priori knowl-edge one incorporates into the models and statistical tests. As birdwatch-ers, we are entirely happy with Diamond claiming that he had picked the two doves a priori to scrutinize. When faced with a plate of a dozen similar-looking species in a tropical bird guide, every birder knows, right off, that few of these species will occur together in the same place. Those that do will usually differ significantly in size or exact habitat. Moreover, there is all the natural history, in the previous chapter, of some species clearly being able to reach some islands yet not occurring there regularly.

Certainly other researchers are unsympathetic, arguing that such natural history knowledge is arbitrary and poorly defined. Perhaps this explains why, to our knowledge, those who criticized Diamond's work have never bothered to either map the distributions of the species or illus-trate the species in question. We have, of course, for this is vital to our un-derstanding. And it is on the combination of that natural history—much of it not easily reducible to zeros and ones in a matrix—with the analy-ses we present here that makes the patterns of species co-occurrences so compelling.

The analysis of pattern in species co-occurrences is now a solved prob-lem. Since 1979, when Connor and Simberloff published their critique of Diamond (1975), critical analyses by many authors coupled with formi-

dable increases in computer power and memory have enabled the creation of software tools to tackle and solve problems that were previously
impossible to solve. Whereas in the past many authors agreed with the
recommendation by Gotelli and Graves (1996) that the sample null communities be unconstrained, or at best weakly constrained, the greater
community of researchers reached a consensus that the number of times
each species occurs and the number of species on each island (the so-
called row and column constraints) should match those of the observed
community. The creation of a uniform random sample null space had
eluded researchers for decades until Miklós and Podani (2004) published
a solution. With the limited computer power available in the past, 1,000
null communities were typically employed to elucidate patterns. Now 1
million or even 10 million sample null communities are used, thus sampling the full null space both uniform randomly and more thoroughly.
Harvey et al. (1983) called into question Connor and Simberloff's (1979)
ensemble metric, stating that ensemble metrics obscured subtle patterns.
We now test and analyze separately every possible species pair. The col-
lective efforts of many researchers from different disciplines and modern
software running on powerful, not to mention portable, computers allow
ecologists to test each and every species pair with a variety of metrics.

Looking back, the road to discovery was not a single superhighway but
rather a more or less biased random walk stumbling toward an unknown
future. Decades ago, so computationally intractable a problem motivated
bizarre solutions, not least among them was Diamond and Gilpin's (1983)
suggestion to create random null communities with fractions to satisfy the
row and column constraints. Because of the difficulty of creating sample
random communities that satisfy the row and column constraints, Connor
and Simberloff (1979) vastly underestimated the sheer size of the sample
null space. We know now that the Vanuatu (formerly the New Hebrides)
sample null space contains no less than 10^{41} unique random communities that satisfy the row and column constraints. We also know that ensemble metrics, those metrics that consist of sums of simpler metrics, obscure subtle patterns by summing and averaging their effectiveness away
(Harvey et al. 1983). As a result of decades of work on both side of the
issues, the study of ecological communities has entered a new future.

We assert that community ecologists now have powerful software and
hardware tools to uncover subtle patterns in nature. The analysis of pattern in species co-occurrences is a solved problem. Much remains to be
discovered, of course. A computer can tell us which species pairs are most
unusual; it remains for us to go forth and explain why.

References

Almeida-Neto, M., P. Guimaraes, P. Guimaraes, Jr., R. Loyola, and W. Ulrich. 2008. A consistent metric for nestedness analysis in ecological systems: Reconciling concept and measurement. Oikos 117:1227–39.

Arbogast, B. S., S. V. Drovetsk, R. L. Curry, P. T. Boag, G. Seutin, P. R. Grant, B. R. Grant, and D. J. Anderson. 2006. Evolution 60(2):370–82.

Atmar, W., and B. D. Patterson. 1994. The measurement of order and disorder in the distribution of species in fragmented habitat. Oecologia 96:373–82.

Baker, R. J., and H. H. Genoways. 1978. Zoogeography of Antillean bats. Academy of Natural Sciences, Special Publication 13:53–97.

Bascompte, J., P. Jordano, C. J. Meila, and J. M. Olesen. 2003. The nested assembly of plant–animal mutualistic networks. Proceedings of the National Academy of Sciences, USA, 100:9383–87.

Bascompte, J., P. Jordano, and J. M. Olesen. 2006. Asymmetric coevolutionary networks facilitate biodiversity maintenance. Science 312:431–33.

Bluthgen, N., F. Menzel, T. Hovestadt, and B. Fiala. 2007. Specialization, constraints, and conflicting interests in mutualistic networks. Current Biology 17:341–46.

Bolger, D. T., C. Alberts, and M. E. Soulé. 1991. Occurrence patterns of bird species in habitat fragments: Sampling, extinction, and nested species subsets. American Naturalist 137:155–66.

Bond, J. 1971. Birds of the West Indies. Collins, London.

Bregulla, H. L. 1992. Birds of Vanuatu. Anthony Nelson, Oswestry, England.

Brualdi, R. A. 1980. Matrices of zeros and ones with fixed row and column sum vectors. Linear Algebra and Its Application 33:159–231.

Brualdi, R. A., and J. G. Sanderson. 1999. Nested species subsets, gaps, and discrepancy. Oecologia 119:256–65.

Carlo, T. A., J. Collazo, and M. J. Groom. 2003. Avian fruit preferences across a Puerto Rican forested landscape: Pattern consistency and implications for seed removal. Oecologia 134:119–31.

Clark, C. E. 1953. An Introduction to Statistics. John Wiley and Sons, New York

Clements, F. E. 1916. Plant Succession. Carnegie Institute Washington Publications 242.

Cody, M. L., and J. M. Diamond, eds. 1975. Ecology and Evolution of Communities. Harvard University Press, Cambridge.

Cohen, J. E., and C. M. Newman. 1985. A stochastic theory of community food webs: I. Models and aggregated data. Proceedings of the Royal Society B 224:421–48.

Cohen, J. E., S. L. Pimm, P. Yodzis, and J. Saldan. 1993. Body sizes of animal predators and animal prey in food webs. Journal of Animal Ecology 62:67–78.

Colwell, R. K., and D. W. Winkler. 1984. A null model for null models in biogeography. In Ecological Communities: Conceptual Issues and the Evidence, ed. D. R. Strong, Jr., D. Simberloff, L. G. Abele, and A. B. Thistle, 344–59. Princeton University Press, Princeton.

Connell, J. H. 1961. The influence of interspecific competition and other factors on the distribution of the barnacle Chthamalus stellatus. Ecology 42:710–23.

———. 1980. Diversity and coevolution of competitors, or the ghost of competition past. Oikos 35:131–38.

Connor, E. F., M. D. Collins, and D. Simberloff. 2013. The checkered history of checkerboard distributions. Ecology 94: 2403–14.

Connor, E. F., and D. Simberloff. 1978. Species number and compositional similarity of the Galápagos flora and avifauna. Ecological Monographs 48:219–48.

———. 1979. The assembly of species communities: Chance or competition? Ecology 60(6):1132–40.

———. 1983. Interspecific competition and species co-occurrence patterns on islands: Null models and the evaluation of evidence. Oikos 41:455–65.

Cook, R. R., and J. F. Quinn. 1995. The importance of colonization in nested species subsets. Oecologia 102:413–24.

Crowell, K. L., and S. L. Pimm. 1976. Competition and niche shifts of mice introduced onto small islands. Oikos 27:251–58.

Curry, R. L. 1986. Whatever happened to the Floreana Mockingbird? Noticias de Galápagos 43:13–15.

Dale, M. R. T. 1984. The continuity of up-slope and down-slope boundaries of species in a zoned community. Oikos 42:92–96.

———. 1986. Overlap and spacing of species' ranges on an environmental gradient. Oikos 47:303–8.

———. 1988. The spacing and intermingling of species boundaries on an environmental gradient. Oikos 53:351–56.

Darwin, C. R. 1859. The Origin of Species by Means of Natural Selection. J. Murray, London.

———. 1862. On the Various Contrivances by Which British and Foreign Orchids Are Fertilised by Insects, and on the Good Effects of Intercrossing. J. Murray, London.

Diamond, J. M. 1975. Assembly of species communities. In Ecology and Evolution of Communities, ed. M. L. Cody and J. M. Diamond, 342–444. Harvard University Press, Cambridge.

Diamond, J. M., and M. E. Gilpin. 1982. Examination of the "null" model of Connor and Simberloff for species co-occurrence on islands. Oecologia 52:64–74.

———. 1983. Biogeographic umbilici and the origin of the Philippine avifauna. Oikos 41:307–21.

Diamond, J. M., and A. G. Marshall. 1976. Origin of the New Hebridean avifauna. Emu 76:187–200.

Diamond, J., S. L. Pimm, M. E. Gilpin, and M. LeCroy. 1989. Rapid evolution of character displacement in myzomelid honeyeaters. American Naturalist 134:675–708.

Dormann C., J. Frund, N. Bluthgen, and B. Gruber. 2009. Indices, graphs and null models: Analyzing bipartite ecological networks. Open Ecology Journal 2:7–24.

Ehrlich, P. R., and P. H. Raven. 1964. Butterflies and plants: A study in coevolution. Evolution 18:586–608.

Elton, C. S. 1966. The Pattern of Animal Communities. Methuen, London.

Forero-Medina, G., J. Terborgh, S. J. Socolar, and S. L. Pimm. Elevational ranges of birds on a tropical montane gradient lag behind warming temperatures. PloS ONE 6, no. 12 (2011): e28535.

Fox, B. J. 1999. The genesis and development of guild assembly rules. In Ecological Assembly Rules, ed. E. Weiher and P. Keddy, 23–57. Cambridge University Press, Cambridge.

Gilpin, M. E., and J. M. Diamond. 1982. Factors contributing to non-randomness in species co-occurrences on islands. Oecologia 52:75–84.

Gleason, H. A. 1926. The individualistic concept of the plant association. Bulletin of the Torrey Botanical Club 53:1–20.

Gotelli, N. J. 2000. Null model analysis of species co-occurrence patterns. Ecology 81(9):2606–21.

———. 2001. Research frontiers in null model analysis. Global Ecology and Biography Letters 10:337–43.

Gotelli, N. J., and L. G. Abele. 1982. Statistical distributions of West Indian land bird families. Journal of Biogeography 9:421–35.

Gotelli, N. J., and G. L. Entsminger. 1999. EcoSim. Null models software for ecology. Version 3.0. Acquired Intelligence Incorporated and Kesey-Bear.

———. 2001. Swap and fill algorithms in null model analysis: Rethinking the knight's tour. Oecologia 129:281–91.

———. 2003. Swap algorithms in null model analysis. Ecology 84(2):532–35.

Gotelli, N. J., and G. R. Graves. 1996. Null Models in Ecology. Smithsonian Press, Washington, DC.

Grant, P. R. 1981. Speciation and the adaptive radiation of Darwin's finches. American Scientist 69:653–64.

————. 1986. Ecology and Evolution of Darwin's Finches. Princeton University Press, Princeton.

Grant, P. R., I. Abbott, D. Schluter, R. L. Curry, and L. K. Abbott. 1985. Variation in the size and shape of Darwin's finches. Biological Journal of the Linnean Society 25:1–29.

Grant, P. R., and D. Schluter. 1984. Interspecific competition inferred from patterns of guild structure. In Ecological Communities: Conceptual Issues and the Evidence, ed. D. R. Strong, Jr, D. Simberloff, L. G. Abele, and A. B. Thistle, 201–33. Princeton University Press, Princeton.

Grant, V., and K. A. Grant. 1965. Flower Pollination in the Phlox Family. Columbia University Press, New York.

Harvey, P. H., R. K. Colwell, J. W. Silvertown, and R. M. May. 1983. Null models in ecology. Annual Review of Ecology and Systematics 14:189–211.

Hofer, U., L.-F. Bersier, and D. Borcard. 1999. Spatial organization of a herpetofauna on an elevational gradient revealed by null model tests. Ecology 80:976–88.

Inger, R. F., and R. K. Colwell. 1977. Organization of contiguous communities of amphibians and reptiles in Thailand. Ecological Monographs 47:229–53.

Inger, R. F., and B. Greenberg. 1966. Ecological and competitive relations among three species of frogs (genus Rana). Ecology 4:746–59.

Ings, T., J. M. Montoya, J. Bascompte, N. Blüthgen, L. Brown, C. F. Dormann, F. Edwards, D. Figueroa, U. Jacob, J. L. Jones, R. B. Lauridsen, M. E. Ledger, H. M. Lewis, J. Olesen, F. J. van Veen, P. H. Warren, and G. Woodward. 2009. Ecological networks: Beyond food webs. Journal of Animal Ecology 78:253–69.

Inouye, D. W., and G. H. Pyke. 1988. Pollination biology in the Snowy Mountains of Australia: Comparisons with montane Colorado, USA. Australian Journal of Ecology 13, no. 2:191–205.

Joppa, L. N., J. Bascompte, J. M. Montoya, R. V. Solé, J. Sanderson, and S. L. Pimm. 2009. Reciprocal specialization in ecological networks. Ecology Letters 12:961–69.

Lack, D. 1947. Darwin's Finches. Cambridge University Press, Cambridge.

————. 1971. Ecological Isolation in Birds. Blackwell Scientific Publications, Oxford.

Lakatos, I. 1978. The Methodology of Scientific Research Programmes: Philosophical Papers. Vol. 1. Ed. John Worrall and Gregory Currie. Cambridge University Press, Cambridge.

Lawton, J. H. 1999. Are there general laws in ecology? Oikos 64:177–92.

LeCroy, M., and F. K. Barker. 2006. A new species of bush-warbler from Bougainville Island and a monophyletic origin for Southwest Pacific Cettia. American Museum Novitates, no. 351.

Levins, R. 1975. Evolution in communities near equilibrium. In Ecology and Evolution of Communities, ed. M. L. Cody and J. M. Diamond, 16–50. Harvard University Press, Cambridge.

MacArthur, R. H. 1958. Population ecology of some warblers of northeastern co-
niferous forests. Ecology 39:599–619.

———. 1972. Geographical Ecology: Patterns in the Distribution of Species.
Harper and Row, New York.

MacKenzie, D. I., J. D. Nichols, J. A. Royle, K. H. Pollock, L. L. Bailey, and J. E.
Hines. 2005. Occupancy Estimation and Modeling: Inferring Patterns and Dy-
namics of Species Occurrence. Elsevier Academic Press, MA.

MacNally, R. C. 1989. The relationship between habitat breadth, habitat position,
and abundance in forest and woodland birds along a continental gradient.
Oikos 54:44–54.

———. 1990. Modelling distributional patterns of woodland birds along a continental gradient. Ecology 71: 360–74.

Manly, B. F. J. 1995. A note on the analysis of species co-occurrences. Ecology
76(4):1109–15.

May, R. M. 1986. The search for patterns in the balance of nature: Advances and
retreats. Ecology 65:1115–26.

———. 1972. Will a large complex system be stable? Nature 238:413–14.

———. 1973. Stability and Complexity in Model Ecosystems. Princeton University
Press, Princeton.

May, R. M., and McLean, A. R. 2007. Theoretical Ecology: Principles and Appli-
cations. Oxford University Press, Oxford.

Mayr, E. 1945. Birds of the Southwest Pacific. Macmillan, New York.

Mayr, E., and J. M. Diamond. 2001. The Avifauna of Northern Melanesia. Oxford
University Press, Oxford.

Menge, B. A. 1995. Indirect effects in marine rocky intertidal interactions webs:
Patterns and importance. Ecological Monographs 65:21–74.

Miklós, I., and J. Podani. 2004. Randomization of presence-absence matrices:
Comments and new algorithms. Ecology 85(1): 86–92.

Montoya, J. M., S. L. Pimm, and R. V. Solé. 2006. Ecological networks and their fra-
gility. Nature 442:259–64.

Moyle, R. G., C. E. Filardi, C. E. Smith, and J. Diamond. 2009. Explosive Pleisto-
cene diversification and hemispheric expansion of a "great speciator." Proceedings of the National Academy of Science, USA, 106:1863–68.

Paine, R. T. 1966. Food web complexity and species diversity. American Naturalist 100:65–75.

Patterson, B. D. 1987. The principle of nested subsets and its implications for biological conservation. Conservation Biology 1:323–34.

Patterson, B. D., and W. Atmar. 1986. Nested subsets and the structure of insular
mammalian faunas and archipelagos. Biological Journal of the Linnean Society
28:65–82.

Patterson, B. D., and J. H. Brown. 1991. Regionally nested patterns of species composition in granivorous rodent assemblages. Journal of Biogeography 18:395–402.

196   REFERENCES

Petren, K., B. R. Grant, and P. R. Grant. 1999. A phylogeny of Darwin's finches based on microsatellite DNA length variation. Proceedings of the Royal Society of London: Biology 266:321–29.

Pielou, E. C. 1975. Ecological Diversity. John Wiley and Sons, New York.

———. 1977. The latitudinal spans of seaweed species and their patterns of overlap. Journal of Biogeography 4:301–11.

———. 1978. Latitudinal overlap of seaweed species: Evidence for quasi-sympatric speciation. Journal of Biogeography 5:227–38.

Pielou, E. C., and R. D. Routledge. 1976. Salt marsh vegetation: Latitudinal gradients in the zonation patterns. Oecologia 24:311–21.

Pimm, S. L. 1982. Food Webs. Chapman and Hall, London.

———. 1986. Putting the species back into community ecology. Trends in Ecology and Evolution 1:51–52.

———. 1991. The Balance of Nature? Ecological Issues in the Conservation of Species and Communities. University of Chicago Press, Chicago.

Pimm, S. L., and C. N. Jenkins. 2010. Extinctions and the practice of preventing them. In Conservation Biology for All, ed. N. S. Sodhi and P. R. Ehrlich. Oxford University Press, Oxford.

Pimm, S. L., C. N. Jenkins, R. Abell, T. M. Brooks, J. L. Gittleman, L. N. Joppa, P. H. Raven, C. M. Roberts, and J. O. Sexton. 2014. The biodiversity of species and their rates of extinction, distribution, and protection. Science 344, no. 6187: 1246752.

Pimm, S. L., and J. H. Lawton. 1980. Are food webs divided into compartments? Journal of Animal Ecology 49:879–98.

Pimm, S. L., and M. L. Rosenzweig. 1981. Competitors and habitat use. Oikos 37:1–6.

Putman, R. J. 1994. Community Ecology. Chapman and Hall, London.

Quammen, D. 1996. The Song of the Dodo: Island Biogeography in the Age of Extinction. Simon and Schuster, New York.

Roberts, A., and L. Stone. 1990. Island-sharing by archipelago species. Oecologia 83:560–67.

Roberts, E. 1986. Thinking Recursively. John Wiley and Sons, New York.

Rosenzweig, M. L. 1981. A theory of habitat selection. Ecology 62:327–35.

Roughgarden, J. 1983. Competition and theory in community ecology. American Naturalist 122:583–601.

Sanderson, J. G. 2000. Testing ecological patterns. American Scientist 88:332–39.

Sanderson, J., J. M. Diamond, and S. L. Pimm. 2009. Pairwise co-existence of Bismarck and Solomon landbird species. Evolutionary Ecology Research 11:771–86.

Sanderson, J. G., M. P. Moulton, and R. G. Selfridge. 1998. Null matrices and the analysis of species' co-occurrences. Oecologia 116:275–83.

Simberloff, D. 1978. Using island biogeographic distributions to determine if colonization is stochastic. American Naturalist 112:713–26.

———. 1983. Competition theory, hypothesis-testing, and other community ecological buzzwords. American Naturalist 122:626–35.

Stone, L., and A. Roberts. 1990. The checkerboard score and species distributions. Oecologia 85:74–79.

———. 1992. Competitive exclusion, or species aggregation? Oecologia 91:419–24.

Strong, D., Jr., L. Szyska, and D. Simberloff. 1979. Tests of community-wide character displacement against null hypotheses. Evolution 33:897–913.

Sulloway, F. J. 1982. Darwin and his finches: The evolution of a legend. Journal of the History of Biology 15(1):1–53.

Temeles, E. J., and J. W. Kress. 2003. Adaptation in a plant–hummingbird association. Science 300:630–33.

Terborgh, T. 1971. Distribution on environmental gradients: Theory and a preliminary interpretation of distributional patterns in the avifauna of the Cordillera Vilcabamba, Peru. Ecology 52:23–40.

———. 1985. The role of ecotones in the distribution of Andean birds. Ecology 66:1237–46.

Terborgh, T., and J. S. Weske. 1975. The role of competition in the distribution of Andean birds. Ecology 56:562–76.

Thébault, E., and C. Fontaine. 2008. Does asymmetric specialization differ between mutualistic and trophic networks? Oikos 117:555–63.

Tonnis, B., P. R. Grant, B. R. Grant, and K. Petren. 2005. Habitat selection and ecological speciation in Galápagos warbler finches (*Certhidea olivacea* and *Certhidea fusca*). Proceedings of the Royal Society of B 272(1565):819–26.

Townsend, C. R., M. Begon, and J. L. Harper. 2003. Essentials of Ecology. Blackwell Publishing, Oxford.

Ulrich W., M. Almeida-Neto, and N. Gotelli. 2009. A consumer's guide to nestedness analysis. Oikos 118:3.

Ulrich, W., and N. J. Gotelli. 2007. Null model analysis of species nestedness patterns. Ecology 88:1824–31.

Underwood, A. J. 1978. The detection of non-random patterns of distribution of species along a gradient. Oecologia 3:317–26.

Vázquez, D. P., and M. A. Aizen. 2004. Asymmetric specialization: A pervasive feature of plant–pollinator interactions. Ecology 85:1251–57.

Verbeek, A., and P. M. Kroonenberg. 1985. A survey of algorithms for exact distribution of test statistics in r by c contingency tables with fixed margins. Computational Statistics and Data Analysis 3:159–85.

Wallace, A. R. 1855. On the law which has regulated the introduction of new species. Annals and Magazine of Natural History, 2nd ser., 16:184–96.

———. 1880. Island life: or, The phenomena and causes of insular faunas and floras, including a revision and attempted solution of the problem of geological climates. Macmillan, London.

Wang, B. Y. 1988. Precise number of (0,1) matrices in A (R,S). Scientia Sinica, ser. A 31:1–6.

Wasserthal, L. T. 1998. Deep flowers for long tongues. Trends in Ecology and Evolution 13:459–60.

Whittaker, R. H. 1967. Gradient analysis of vegetation. Biological Reviews of the Cambridge Philosophical Society 42:207–64.

Whittaker, R. H., and W. A. Niering. 1965. Vegetation of the Santa Catalina Mountains, Arizona: A gradient analysis of the south slope. Ecology 46:429–52.

———. 1975. Vegetation of the Santa Catalina Mountains, Arizona. V. Biomass, production, and diversity along the elevational gradient. Ecology 56:771–90.

Wilson, J. B. 1987. Methods for detecting non-randomness in species co-occurrence: A contribution. Oecologia 73:579–82.

Wright, D. H., B. D. Patterson, G. M. Mikkelson, A. Cutle, and W. Atmar. 1997. A comparative analysis of nested subset patterns of species composition. Oecologia 113:1–20.

Wright, D. H., and J. H. Reeves. 1992. On the meaning and measurement of nestedness of species assemblages. Oecologia 92:416–28.

Wright, S. J., and C. C. Biehl. 1982. Island biogeographic distributions: Testing for random, regular, and aggregated patterns of species occurrence. American Naturalist 119(3):345–57.

Zaman, A., and D. Simberloff. 2002. Random binary matrices in biogeographical ecology: Instituting a good neighbor policy. Environmental and Ecological Statistics 9:405–21.

Zar, J. H. 1999. Biostatistical Analysis. Prentice Hall, Englewood Cliffs, NJ.

Index

Page numbers in italics refer to figures.

204

INDEX

Nectarinia sericea, 148, 157
network, 174–77, 180, 181, 183–85
    antagonistic, 181
    daughter, 183
    mutualistic, 181, 183
    subnetwork, 183
    trophic, 184, 185
New Guinea, 7, 8, 16, 17, 20, 24, 35
    fruit doves, 19
    West Papua, 23
niche
    actual, 40
    compression, 40
    potential, 40
Niering, W., 161
night sky effect, 4
nonrandom, 57
null space, 67, 70, 71, 97, 105, 106
    full, 57, 61, 62, 70, 73, 75, 76, 78, 79, 84,
        85, 102
    sample, 9, 50, 61, 62, 70, 73, 74, 75, 84, 87,
        90, 91, 92, 109, 110, 129, 138

orchid, 179
    Malagasy, 178
ordering, canonical, 77
Origin of Species, The (Darwin), 132, 135
overexploitation, 40, 41

Pachycephala, 153, 158
    melanura, 36
    pectoralis, 36
Paine, R., 12, 34
palimpsests, 181, 185
pattern, 28
    biogeographical, 14
    checkerboard, 14
    nature, 8, 10, 11
    nestedness, 14, 170, 171, 173, 176, 177,
        184, 185, 188, 189
    observed, 33
    parapatric, 160
    reciprocal specialization, 15, 170, 174,
        177, 178–81, 183–85, 188, 189
Pattern of Animal Communities, The
    (Elton), 5
Patterson, B., 14, 63, 173, 177
Petren, K., 141
Petroica multicolor, 118
Philippines, 20

Phylloscopus amoenus, 155
Pielou, E., 63, 162, 167
pigeons, 20
Pimm, S., 3, 5, 9, 55, 170, 171, 188
Pinus edulis, 160
Pinyon-juniper woodland, 160
Pisaster, 34
Platyspiza crassirostris, 140
Podani, J., 86, 87, 90, 91, 190
pollinator, 177–79
population, sympatric, 14
Pratt, D., xi
predation, 12, 34
primates, 3
Princeton University, x, 5, 17
probability density function
    distribution-free, 108, 138
    nonparametric, 108
Ptilinopus, 18, 19, 20, 26, 53, 153
    coronulatus, 20
    distribution, 22
    greyi, 19, 20, 53
    insolitus, 19, 20, 22, 23, 28, 29
    iozonus, 20
    pulchellus, 20
    rivoli, 19, 20, 23, 36, 42
    solomonensis, 19, 20, 23, 26, 28, 29, 36,
        42, 150
    superbus, 19, 20, 26, 29, 150
    tannensis, 53
    viridis, 19, 20, 23
Putnam, R., ix
Pyke, G., 183

quasi-swap, 88, 89

random, 6, 7, 8, 42, 45, 53, 57, 58, 63, 64, 66
    uniform, 84, 85, 86, 90
range, geographical, 3
Raven, P., 178
Reinwardtoena browni, 36, 37, 39
representative, 70, 73, 74
resource
    partition, 43
    utilization, 41
Rhipidura, 53, 153
    dahli, 156
    leucophrys, 156
    rufifrons, 156
Roberts, A., 79, 94, 98, 100–106, 108, 113